Studies in Economic Ethics and Philosophy

Springer
Berlin
Heidelberg
New York
Barcelona
Hong Kong
London
Milan
Paris
Singapore
Tokyo

Studies in Economic Ethics and Philosophy

P. Koslowski (Ed.)
Ethics in Economics, Business, and Economic Policy (out of print)
X, 178 pages. 1992
ISBN 3-540-55359-2

P. Koslowski and Y. Shionoya (Eds.)
The Good and the Economical
Ethical Choices in Economics and Management (out of print)
X, 202 pages. 1993
ISBN 3-540-57339-9

H. De Geer (Ed.)
Business Ethics in Progress?
IX, 124 pages. 1994
ISBN 3-540-57758-0

P. Koslowski (Ed.)
The Theory of Ethical Economy in the Historical School
XI, 343 pages. 1995, reprinted 1997
ISBN 3-540-59070-6

A. Argandoña (Ed.)
The Ethical Dimension of Financial Institutions and Markets
XI, 263 pages. 1995
ISBN 3-540-59209-1

G.K. Becker (Ed.)
Ethics in Business and Society.
Chinese and Western Perspectives
VIII, 233 pages. 1996
ISBN 3-540-60773-0

P. Koslowski
Ethics of Capitalism and Critique of Sociobiology. Two Essays with a Comment by James M. Buchanan
IX, 142 pages. 1996
ISBN 3-540-61035-9

F. Neil Brady (Ed.)
Ethical Universals in International Business
X, 246 pages. 1996
ISBN 3-540-61588-1

P. Koslowski and A. Føllesdal (Eds.)
Restructuring the Welfare State
Theory and Reform of Social Policy
VIII, 402 pages. 1997
ISBN 3-540-62035-4

G. Erreygers and T. Vandevelde
Is Inheritance Legitimate?
Ethical and Economic Aspects of Wealth Transfers
X, 236 pages. 1997
ISBN 3-540-62725-1

P. Koslowski (Ed.)
Business Ethics in East Central Europe
XII, 151 pages. 1997
ISBN 3-540-63367-X

P. Koslowski (Ed.)
Methodology of the Social Sciences, Ethics, and Economics in the Newer Historical School
From Max Weber and Rickert to Sombart and Rothacker
XII, 565 pages. 1997
ISBN 3-540-63458-4

A. Føllesdal and P. Koslowski (Eds.)
Democracy and the European Union
X, 309 pages. 1998
ISBN 3-540-63457-6

P. Koslowski (Ed.)
The Social Market Economy
Theory and Ethics of the Economic Order
XII, 360 pages. 1998
ISBN 3-540-64043-6

Amitai Etzioni
Essays in Socio-Economics
XII, 182 pages. 1999
ISBN 3-540-64466-0

Peter Koslowski (Ed.)

Sociobiology and Bioeconomics

The Theory of Evolution in Biological and Economic Theory

With 18 Figures
and 1 Table

Springer

Prof. Dr. Dr. h.c. Peter Koslowski
Hannover Institute of Philosophical Research –
Forschungsinstitut für Philosophie Hannover
Centre for Ethical Economy and Business Culture
Gerberstrasse 26
D-30169 Hannover
Germany

Published with the Support of
STIFTUNG FORSCHUNGSINSTITUT FÜR PHILOSOPHIE HANNOVER

ISBN 3-540-65380-5 Springer-Verlag Berlin Heidelberg New York

Cataloging-in-Publication Data applied for
Die Deutsche Bibliothek – CIP-Einheitsaufnahme
Sociobiology and bioeconomics: the theory of evolution in biological and economic theory / Peter Koslowski (ed.). – Berlin; Heidelberg; New York; Barcelona; Hong Kong; Milan; London; Paris; Singapore; Tokyo: Springer, 1999
(Studies in economic ethics and philosophy)
ISBN 3-540-65380-5

Printed in Germany

Hardcover Design: Erich Kirchner, Heidelberg

SPIN 10701315 42/2202-5 4 3 2 1 0 – Printed on acid-free paper

Preface

The volume at hand publishes the proceedings of the conference "Developmental Systems, Competition, and Cooperation in Sociobiology and Economics", a "Studies in Economic Ethics and Philosophy (SEEP)-Conference" held at Marienrode Monastery, Hildesheim-Marienrode near Hannover, Germany, on 24-28 April 1996. This "Studies in Economic Ethics and Philosophy (SEEP)-Conference" was made possible by the support of the Stiftung Forschungsinstitut für Philosophie Hannover – Foundation Hannover Institute of Philosophical Research.

The editor wishes to thank his co-workers at the Forschungsinstitut für Philosophie Hannover for their assistance in organizing the conference at Marienrode and Anna Maria Hauk M.A. and Bettina Löhnert M.A. for their assistance in preparing the computerized version of this book.

A special word of gratitude is due the Stiftung Forschungsinstitut für Philosophie Hannover whose financial support made the conference in 1996 possible.

Hannover, September 1998 P. K.

Contents

CONTENTS

Part II

Natural Selection and Developmental Systems: Redrawing the Boundaries Between Genetic and Non-Genetic Factors of Evolution

Part III

Redrawing the Boundaries Between the Narratives of Evolutionary Biology and the Narratives of Hermeneutics, Mythology, and Philosophy

CONTENTS

Sociobiology, Theory of Evolution, and Bioeconomics
Introduction

PETER KOSLOWSKI

To the Memory of
Reinhard Löw (1949-1994),
Outstanding Scholar of Naturphilosophie

Sociobiology forms a unifying theory of the social interactions in the human and in the animal world. Its principle unifying human and animal societies is the maximisation of genetic fitness by the individual. The maximisation of genetic fitness forms the basic "teleology" of the individual in society. As an application of the Neo-Darwinian Synthesis of evolution theory sociobiology extends Darwinian evolutionary theory to the realm of the social and economic sciences. Sociobiology becomes bioeconomics or the analysis of the economy of nature since individuals and species compete for the same scarce resources of nature to maximise their inclusive fitness and the survival of their off-spring. To make economic use of the resources of nature is a necessary part of maximising inclusive fitness. The life sciences are led by their own research to economic theory for being able to analyse the economy of nature. To live means to make economic use of one's own resources and of those of one's environment. Wasting resources implies reducing the development of life forms.

The acknowledgment of the economy of nature and the importance of economic laws in the natural and in the social evolution as well as in their co-evolution is a feature shared by the Neo-Darwinian Synthesis and evolutionism on the one hand and the older natural right tradition on the other. The laws of nature and of its economy form constraints on the variations of human culture that can be extended and modified but can be done so only with "bio-economic" costs. Sociobiolgy and bioeconomics can make us aware of these costs and support a rational behaviour towards nature's constraints. The extension of biological theory and of the life sciences to the human society and economy is not in itself illegitimate. Rather, it adds important new insights into the nature and constraints of human societal organisation.

Sociobiology as a general field of analysis is therefore of interest to the life sciences and to social theory and philosophy. Sociobiology is, however, used in a twofold way as the field of analysis of human and animal societies and as a particular theory in this field extending the Darwinian evolutionary theory to a unified metaphysics and a total theory of being and of the universe. The latter approach introduced by E.O. Wilson renders sociobiology into an ontology and metaphysics of all being. By doing so, sociobiology or evolutionism as an ontological or metaphysical theory must defend its claim to metaphysics with metaphysical or philosophical arguments. The "metaphysical" sociobiologists, however, do not accept this philosophical task. It is not taken up by authors like Dawkins and fulfilled only in broad outline by E.O. Wilson.

The philosophical discussion and criticism of the ontological and metaphysical assertions of evolutionism and sociobiology have been conducted in the past by the group CIVITAS in three volumes of collections of essays in the German language.[1] The volume at hand is intended to make further progress in this debate and to extend it to new frontiers of inquiry.

---

1 The three CIVITAS volumes are: 1. PETER KOSLOWSKI, PHILIPP KREUZER, REINHARD LÖW (Eds.): *Evolution und Freiheit. Zum Spannungsverhältnis von Naturgeschichte und Mensch* (Evolution and Freedom. On the Tension Between the History of Nature and Humankind), Stuttgart (S. Hirzel Verlag) 1984 (= CIVITAS Resultate Vol. 5). Japanese translation Tokyo 1991.
2. ROBERT SPAEMANN, PETER KOSLOWSKI, REINHARD LÖW (Eds.): *Evolution und menschliches Selbstverständnis* (Evolution Theory and Human Self-Understanding), Weinheim (Acta humaniora) 1984 (= CIVITAS Resultate Vol. 6).
3. ROBERT SPAEMANN, REINHARD LÖW, PETER KOSLOWSKI (Eds.): *Evolutionismus und Christentum* (Evolutionism and Christianity), Weinheim (VCH/Acta humaniora) 1986 (= CIVITAS Resultate Vol. 9).
My own contribution to the first volume has been further elaborated in P. KOSLOWSKI: *Evolution und Gesellschaft. Eine Auseinandersetzung mit der Soziobiologie*, Tübingen (Mohr Siebeck) 1984, 2nd edition 1989. English translation: *Ethics of Capitalism and Critique of Sociobiology*. Two Essays with a Comment by James M. Buchanan, Berlin, New York, Tokyo (Springer) 1996 (Studies in Economic Ethics and Philosophy). - Central is also R. SPAE-

INTRODUCTION

The volume at hand continues not only the philosophical debate on the Neo-Darwinian Synthesis but takes up specific points of crisis within the Darwinian approach. It discusses the points of theoretical crisis and the missing links of Darwinism. It examines the question whether the second law of thermodynamics can be reconciled with the premise of Darwinism that natural selection leads to differentiation and higher complexity of forms of life. The book investigates the question how genetic-somatic and cultural-social systems of heredity interact and how interacting somatic and cultural-mental systems of heredity could be conceptualised in developmental systems of biological and cultural evolution. It analyses the interaction of genetic and non-genetic factors of evolution and redraws the boundaries between biological and hermeneutic interpretations of the genetic and biological code as well as the boundaries between the narratives of evolutionism and the narratives of mythology and philosophy. In contrast to the former three CIVITAS volumes that centred around the debate between philosophers and biologists, this volume develops the debate on the points of crisis of Darwinism in a discussion amongst biologists and social scientists and in a discourse about new directions of research within biology.

The other, new contribution of this volume to the debate on a unified theory of the life sciences and the social sciences is the synthesis of evolutionary biology and evolutionary economics, of biological and economic theory. The economy of nature forms the starting point for an integrated bioeconomic analysis of symbiosis, mutualism, competition, and co-operation as well as of the conditions under which they arise and become evolutionary stable strategies. What is the comparative role that phenomena like symbiosis, mutualism and co-operation play in animal societies and in human societies and economies? Are there common features of competition and co-operation in animal and human societies? Finally, the relationship between market selection and natural selection is examined.

The merger of biological and economic analysis in bioeconomics is also relevant for questions of the environment and of ecological equilibrium as well as for theories of economic development. What are the selective biological constraints on economic development, what the developmental constraints on natural selection? The ecological-economic debate about sustainable deve-

---

MANN, R. LÖW: *Die Frage Wozu. Geschichte und Wiederentdeckung des teleologischen Denkens* (The Question What for? History and Rediscovery of Teleological Thinking), München (Piper) 1981, several editions.

lopment and the question whether the economy should be conceptualised in mechanistic or in organismic terms is part of the rising new field of bioeconomics.

Part I

Neo-Darwinian Synthesis, Sociobiology, and the Economy of Nature

Chapter 1

Darwinian Monism: The Economy of Nature

MICHAEL T. GHISELIN

Late in September of 1838 Charles Darwin read a book by Robert Malthus, and discovered natural selection. Darwin, then 29 years old, was the Secretary of the Geological Society of London. Malthus, who had died two years previously, had been the first professor of economics. The interdisciplinary relationship may serve to remind us of two very important points. First, Darwin transformed biology into an historical science, in the sense that geology is an historical science. Second, Darwin transformed biology into an economic science, by showing that the natural economy and the political economy are variations upon a common theme (Hirshleifer 1978).

MICHAEL T. GHISELIN

I. Bioeconomics

When Darwin read Malthus he came to realize that evolutionary change occurs because of a special kind of competition – reproductive competition – between organisms of the same species. Given that organisms vary indefinitely, and given that there is some tendency for like to beget like, successive generations will accumulate those properties that endow their possessors with whatever it is that makes for reproductive success. So life as it exists at any point in time is the product of a long history, and the organisms do whatever it was (as a matter of contingent fact) that the most successful reproducers did. To understand what life is all about we need to know its history. But in addition to the contingent historical facts we need to know the laws of nature that govern such goings on. And these laws are, among other things, the laws of economics.

Economics, broadly defined, is the science of resources, and is sometimes called "general" or "universal" economics to emphasize its "monistic" character (Radnitzky 1992). An emerging interdisciplinary field, bioeconomics, seeks to make the connections clear and explicit (Ghiselin 1992). Bioeconomics looks upon organisms as capital goods, the business of which is producing more such capital goods. Their reproductive success depends upon how they allocate their resources: and laws about diminishing returns and the like tell us how such allocation contributes to reproductive success. (E.g. when the time has come to stop growing and start to reproduce.) The use of economic language has become increasingly common during the past two decades among biologists interested in various aspects of resource allocation, matters with obvious parallels in consumer choice theory and the like. Biologists, however, have largely overlooked what the economists have had to offer, and economists have not taken much advantage of the opportunities either. I must emphasize that bioeconomics, like biochemistry and biophysics, is a hard core natural science, one that achieves its unification of knowledge by subsuming the particular under the more general (just as chemists subsume "bromine" and "iodine" under "halogen"). It is not a matter of figures of speech ("metaphors").

Organisms reproduce, and because they reproduce differentially, species evolve. Species are populations that constitute lineages, which can become transformed indefinitely and still remain the same species. That is one kind of evolution. In addition, species can also proliferate, breaking up and turning into groups of species, or separate lineages. That is another kind of evolution: speciation. Darwin realized that when speciation occurs repeatedly the lineages thus generated tend to diversify, and to occupy different places in the economy of nature. This was a rather late elaboration of his theory, and again was stimulated by economic considerations. He was strongly influenced by the work of the French physiologist Henri Milne Edwards, who applied the principle of the division of labor to animal physiology (e.g. Milne Edwards 1851). Specialists are often at a competitive advantage, because they make more effective use of resources. So competition within the natural economy brings about diversification, and by extension all sorts of other complex relationships among organisms and among species. Thus, further insight about economics led Darwin to found the modern science of ecology. One branch of ecology that Darwin founded was pollination ecology, with its countless examples of mutualism and reciprocity. Cooperation between organisms gives rise to an economic unit that allows its participants to compete more effectively. In some cases, but not all, specialists and cooperators actually utilize resources more effectively, and the result may be an increase in the level of economic activity.

II. Individuality

This view of life, to use Darwin's expression, entailed a radical departure from metaphysical tradition. Implicitly, at least, it rejected the notion that biological species are kinds of organisms – or kinds of anything whatsoever. Modern evolutionary biology, at least during the last two decades, has come to do so explicitly (Ghiselin 1974a, 1997). If species were "kinds" that would mean they were classes – hence abstractions. But abstract classes – such as "the chair" – cannot do anything. We can sit on a concrete, particular chair, but not upon "chair" in the abstract. So too with biological species: they are concrete individuals like ourselves, and like the chairs upon which we sit. So species are supraorganismal wholes, of which the organisms are parts. And the families, orders and other "higher taxa" are likewise concrete particulars: branches of genealogical trees. The living world is a whole-part hierarchy, or rather a host of them.

The implications may not be immediately obvious, but they are momentous. For one thing, if species and other taxa are indeed individuals, they do not have what are called "essences" or "natures". When people speak of an "essence" or a "nature" they mean, at a minimum, the defining properties of a class. And often some kind of "standard" or norm is implied. Individuals don't have such things, except in the misleading sense of having the properties that are defining of the classes of which they happen to be members. Individuals can be described, but not defined. They are variable, they can change indefinitely, and what properties they have is a matter of contingent fact: they could have been otherwise and they may yet be otherwise.

What, then, does Darwinism tell us about "human nature"? It tells us that there is *no such thing*. Human nature is a metaphysical delusion. Over and above being part of *Homo sapiens* however, nothing, absolutely nothing whatsoever, makes an organism a human being. Yes, we have a history, and yes, we are burdened with its consequences (such as backaches and hernias). But that history could have been very different indeed: *Homo sapiens* would still be *Homo sapiens* if our ancestors had not shifted to bipedal locomotion or done any of those other things that we consider particularly interesting and important about ourselves.

III. Teleology

Because of such historicity, the Darwinian theory renders it manifest that the teleological view of the living world is likewise a metaphysical delusion. Although things are changing, and going from one place to another, they are not going anywhere in particular. The world came into being without planning, foresight, or purpose. The notion of a final cause – that one thing exists for the sake of another – has no legitimate place in Darwinian thinking.

Once we have made this point, however, we had better emphasize that nothing in Darwinian biology rules out such things as intentions and functions. Organisms with foresight and understanding of course may have greater reproductive success than those with less of such things. And the fact that different organs often engage in different activities that have to do with making a living and producing offspring is exactly what one would expect – given the economic advantages to the division of labor. So when we ask what the function of something is, we are not asking about "purpose" or any other occult property, but addressing the issue of how labor is divided. And when we talk about adaptation, we are discussing what has happened, and leave consideration of what is "good" to the mystics. Evolutionary biology is an historical science, and its explanations consist of historical narratives.

IV. Laws

One more point needs to be made about the individuality of taxonomic groups. Such groups are indeed, historical entities, with beginnings and ends, and with definite positions in space and time. The laws of nature make no reference to any individual as such. They are "about" classes of individuals. In the physical sciences there are no laws for such individuals as the Earth, Sol, or the Milky Way. In biology there are no laws for Julius Caesar, General Motors, *Homo sapiens* or Chordata. The laws of nature are about kinds of individuals: such as all celestial bodies irrespective of place and time. Astronomy gives the individuals, astrophysics gives the laws. In evolutionary biology, the laws of nature are also about kinds of individuals: kinds of organisms, kinds of species (e.g. inbred or outbred) and the like. By way of rough approximation, in bioeconomics systematics gives the individuals and economics gives the laws. Many pre-Darwinian biologists expected to explain the taxonomic hierarchy of groups within groups in the same way that chemists have in fact explained the properties of classes of chemicals: by means of laws of nature. Subsequently anti-evolutionists have repeatedly sought to exhume the corpse of such a notion, but they have singularly failed.

V. Hierarchy

Understanding the living world as a whole-part hierarchy helps us to clarify the roles of the individuals that occupy different "levels" in that hierarchy (or complex of hierarchies). They participate in somewhat different processes. Organisms procreate, and species speciate, but genera and higher taxa don't do anything. Let us consider what this means in a simple and straight-forward case. Organisms are the units that do or do not have reproductive success, and therefore may reasonably be said to be selected. They do so by virtue of the collective economic consequences of the properties of their components. The parts of the organism do not stand by themselves, but are reproduced in proportion as the whole reproduces. Parts of organisms are not selected; rather they affect and are affected by selection. Both the organism and the parts are adapted but the adaptation of the parts is a contextual matter. It is organisms, not their genitalia, that copulate. To say that natural selection "acts upon" either the organisms or any of their parts is nonsense, like saying that copulation "acts upon" something. It can't, because it isn't an agent (see Ghiselin 1981).

Turning now to species and other populations, it is surely they, rather than their component organisms, that evolve. But when we say that species evolve, we ordinarily mean that the component organisms have different properties from the ones that they have replaced. If the evolutionary change in question is due to natural selection, then the product of this process is a species composed of adapted organisms, and not an adapted species. It is a matter of the organisms, not the species, having differential reproductive success. This is why evolutionary biologists do not invoke "species-level" or "population-level" adaptations under ordinary conditions. The basic cause is competition between organisms within the same species.

However, and this is very important, it is perfectly possible for supraorganismal individuals to compete as units, and to reproduce differentially. Darwin explained the evolution of "sterile castes" in social insects by treating the family as a supra-organismal individual. After all, multicellular organisms, with their somatic and germ cells, are a perfectly good precedent. In order for that kind of thing to happen, however, the lower level individuals have to be "held together" in a way that makes them a "cohesive" whole. Such restrictions make it methodologically appropriate to treat "group selection" as less plausible than the straight forward organismal mode. Be this as it may, the issues are empirical. In my own writings on that subject I have always opposed group selectionist explanations on the basis of fact – not on the basis of methodological rules turned into metaphysical posits.

Because species are individuals, it is at least possible for them to proliferate differentially, so there can be a kind of selection among them. If some species speciate more than others, or if species differ in their extinction rates, then one would expect at least some cumulative effects. I have acknowledged this possibility from the outset, and have not been able to rule it out. However my own empirical investigations have led me to de-emphasize it (Ghiselin 1974b). On the other hand Niles Eldredge has taken species selection very seriously and in so doing he and others have given us a much better picture of macroevolution (see Eldredge 1985, 1989). His investigations have included some consideration of economics, though I am not sure that we agree altogether about what economics is (Eldredge 1986). At any rate it must be emphasized that such models by no means have to contradict the fundamental principles of Darwinian biology.

This suggests that there is a lot more to Darwinism than just organismal reproductive competition. In fact, Darwin repeatedly stated that he considered natural selection to be the main, but not exclusive, means of modification. What were these additional causes of modification? Sexual selection, for one. But artificial, natural, and sexual selection are all variants upon one grand theme, not something fundamentally different. Darwin also attributed a minor role to certain "Lamarckian" mechanisms that we now reject as implausible. We do accept a certain amount of genetic drift, but that just adds another stochastic element, something that is hardly out of line with Darwin's basic theory.

VI. Development

When organisms reproduce, they replicate both the whole and the parts. But such replication is by no means perfect. Darwin spoke of "variation" pure and simple. These days the terms "mutation" and "recombination" are more commonly used because they describe what goes on at the chromosomal level, which is were the action takes place. The basic point, however, remains the same: the reproductive and developmental apparatus generates differences among the organisms within a species. And the course of evolution is affected by the properties of that very apparatus.

Contrary to what has so often been asserted, Darwin did not claim that variation is "random" in the sense that one variation is just as probable as another. Rather, he said that it is "fortuitous" which is quite a different matter. One of Darwin's non-selective mechanisms of change was "correlated variation". If one selects for whiteness in blue-eyed cats, one gets a breed of cats that are deaf as well; the three properties are causally interrelated. So the mechanics of variation, and the mechanics of development, are important influences upon the course of evolutionary history.

In a very fundamental sense, Darwin considered evolution to be the outcome of changes in the manner in which organisms develop. Unfortunately his ideas upon such matters have been widely suppressed by his opponents (notably William Bateson, Richard Goldschmidt, and Stephen Jay Gould). The reason is not far to seek. Embryology has always been the last refuge of the anti-evolutionist.

If one treats evolution as if it were like the development of an embryo, and therefore an "ontogeny writ large", then provision is made for a kind of "design on the installment plan" as John Dewey (1910) characterized it. Another alternative is to treat evolution as if it were like the formation of a crystal, hence the result of laws of nature, perhaps divinely ordained. Evolution is then no longer a matter of "chance" (i.e. history) but of "law" (Hertwig 1922). Ignoring the mystical and ideological appeal of such "orthogeneticist" versions of what is misleadingly called "evolution", we need only dismiss the so-called "biology" as bad metaphysics. Evolution is not some kind of change from a state of potentiality to a state of actuality such as what we observe in the development of an embryo. The developmental apparatus itself gets restructured through time. Surely, there are laws of nature that govern development, and these are not just the laws of chemistry and physics. But the developmental processes that give rise to organisms are largely a matter of contingent, historical fact, and future progress in embryology will consist mainly of working out the phylogeny of development.

In spite of pretentious claims about new evolutionary paradigms somehow grounded in embryology or perhaps in physico-chemical forces, we see little evidence that such hopes are being realized. Instead, molecular biology has become a valuable new tool for working out phylogenetic trees, and developmental genetics has likewise become historical. We are no longer limited to observations upon cells, organs, and organ systems when we reconstruct the history of the body. We have the capacity to identify the genes that control development, to compare them in diverse lineages, and thereby to reconstruct their history as well. The result is an evolutionary physiological anatomy that is very historical, and very Darwinian in spirit.

The term "neo-Darwinism" originally meant the position that natural selection is not just the main, but the exclusive, cause of modification. I don't know of any really good examples. Certainly August Weismann, who is supposed to have founded the neo-Darwinian movement, hardly qualifies, for he accepted both sexual selection and correlated variability. The term has tended to be associated with the rejection of Lamarckian notions and various other alternatives to natural selection. This helps to explain why the Synthetic Theory has been characterized as "neo-Darwinian". Beyond such loose usage, "neo-Darwinian" is a pretty vacuous term.

VII. Genetics

Because the Synthetic Theory was strongly influenced by a certain kind of genetics one may easily get the false impression that evolutionary biologists have some kind of consensus about what genes are, what they do, and why they are important. No such consensus exists. Be this as it may, the notion that genes are the fundamental reality and that organisms and species are somehow epiphenomenal, is a kind of ontological reductionism that is endorsed by a small minority of biologists, mainly persons who are interested in certain aspects of social behavior, and who sometimes call themselves sociobiologists. A lot of us who have made substantive contributions to the scientific study of the phenomena in question prefer not to accept that label, because we reject the ideology, the theology, and the metaphysical kitsch with which it has been burdened.

Calling a gene a "replicator" was a clever piece of sophistry (Dawkins 1976). The suffix "-or" in English as in Latin implies an active agent: that which produces copies, not the passive patient of which copies are made. By such verbal ploys it is fairly easy to endow the part with what is in reality the property of the whole, and even to deny the very existence of the whole. Or, by invoking something the very existence of which evolutionary biology categorically denies, genes are said to be the "final cause" of the rest of the body (Cronin 1991). Furthermore, some of Dawkins's "alleles" are chromosomal deletions: missing segments of DNA the absence of which may be favored by selection. How can something that does not exist be "selfish"? Dawkins (1982, p. 164) acknowledges that there can be an economic advantage to such deletion. But rather than acknowledging the obvious *reductio ad absurdum*, he gives us a *credo quia absurdum*: the absence is a replicator, favored by selection. Next thing we know he will be marketing abridged editions of his books on the ground that these contain more "memes" than the originals! I am afraid that a lot of non-biologists, including some economists, have the false impression that such bad metaphysics is good biology (Elworthy 1993). *Caveat emptor*!

In real evolutionary biology, genes of course are very important, and not just because keeping track of them is a fruitful way to do research. They are largely responsible for each generation of organisms being pretty much like the previous generation. And they play a crucial role in development. From the point of view of bioeconomics we may regard genes as very important too, for they are resources that allow organisms to do such things. But like many other resources they deteriorate, and become obsolete; they often need to be modified, and replaced by something appropriate to new circumstances. By means of sex, organisms overcome the conservative tendencies of the genetic material that of course have some advantages in the short run.

At present evolutionary biologists have pretty much given up trying to come up with a fully satisfactory explanation for the existence of sex (see articles in Michod/Levin 1988). Part of the explanation, at least, would seem to be that as genes are replicated, change due to mutation is inevitable. Sex is one way of getting rid of the defective genetic material. There also seems to be some economic advantage to generating variants. In spite of what the reductionist metaphysics of sociobiology tells us, organisms go to considerable expense getting rid of genes and replacing them with other ones. Sex is probably a "supra-organismal adaptation" only insofar as it need not happen every generation for its advantages to be realized. In other words, it is not there, as an older generation put it, for the good of the population. However, there are some economic consequences of sex that are very important at supra-organismal levels. Were it not for sex, there would be no species, and species provide a kind of supra-organismal repository of whatever it is that genes may provide. Or, if you prefer, species provide something like a market in which genetical resources can be exchanged.

VIII. Culture

Organisms also acquire valuable resources when they learn something about their environments. Such "knowledge" (in a broad sense) can be stored, shared with other organisms, and passed on to later generations. There has been no dearth of efforts at creating a sort of dual-inheritance approach to evolution, with both a genetical and a cultural aspect. Some of these, especially the ones with strong bioeconomic components, have been remarkably successful (Boyd/Richardson 1985; Durham 1991). From the monistic point of view, however, it is important to stress that such resources do not exist in isolation. A new way of coping with the economic problems of life can bring a very different assemblage of selection pressures into play. Learning to use a new kind of food might bring about just a few minor changes in digestive enzymes if any. On the other hand, major changes might also result, especially if the animal comes to forage in a different habitat – with different predators and the like. Technology of course has brought about such shifts, relaxing some selection pressures, intensifying others, perhaps bringing about new ones. We can be confident that other elements of culture have done the same thing.

A culture, like a gene-pool, is a supraorganismal whole, one that provides resources to the organisms that participate in it. In bioeconomic terms, again, we may regard organisms as capital goods, the business of which is producing other capital goods. They accumulate capital and transfer it to subsequent generations, partly by sharing and exchanging resources with other organisms. The economist Gary Becker (1975) refers to the "human capital" that we accumulate and invest in the young, as by feeding, clothing, and educating them; to a bioeconomist this is a particular case of what might be called "organismal capital" that can be turned into a kind of endowment for an organism's offspring. It is indifferent whether such capital investment is provided in the form of material goods, as when a doe provides milk to her fawn, or in the form of services, as when an ostrich cares for his chicks. Likewise, education, whether formal or not, is a transfer of epistemic resources, ones that are particularly important where life is as complicated as our own.

IX. Morality

When economics is misconstrued in terms of crass materialism and unenlightened self-interest, one readily forgets that man does not live by bread alone. Economic theory takes our wants for granted, and considers how we might allocate our resources in satisfying those wants. It does not tell us that our appetites for goodness, truth, and beauty do not exist, nor does it tell us whether we should indulge them. Evolutionary theory, however, has been somewhat of a loss to find a straight-forward explanation for such phenomena. There is no reason, however, why the explanation has to be straight-forward, or simplistic. When a society emerges wherein the organisms are profoundly aware of what is going on around them, wherein they can consider alternatives to their present condition of existence, and wherein they can join with one another to alter their circumstances, all sorts of new possibilities are opened up. For instance, the discovery of justice, and the institution of law. If things like that happen, then the selection pressures might be profoundly altered.

Darwin was very interested in the origin of what Adam Smith had called "the moral sentiments". Like later workers he sought for indications of protoethical behavior in parental care and in the life of the hive. He suggested that if we were like bees, in which the workers kill their brothers the drones, fratricide would be a matter of doing one's duty and following one's conscience (Darwin 1871). It would be a matter of sacrificing another's life so that the family as a whole would enjoy a competitive advantage over conspecific families. Sacrificing one's own life for the same basic reason would be just a variant upon the same theme. Among the various explanations that he entertained for morality Darwin favored treating such higher-level individuals as the tribe as competitive units. He explained how selection works in this case in spite of the death of the self-sacrificing organism. When a breeder of domesticated animals or plants finds a desirable trait in a sterile organism or a dead one, the trick is to "breed from the same stock" – relatives are apt to have the same trait. This is the basic principle of "kin selection" as it is now called.

People who study the behavior of social insects have been particularly enamored of kin selection and the "inclusive fitness" in terms of which it is discussed. To those whose metaphysical preferences have a reductionist character, gene-gene competition obviously has greater appeal than does organism-organism or family-family competition. Rather than saying that families, the components of which had certain properties, have outreproduced families, the components of which had other properties, it is said that the components of those families "maximize their inclusive fitness". On the one hand attention is diverted from what is going on at higher integrational levels. On the other hand a propensity, that is to say, an abstract dispositional property, called "fitness", substitutes for a causal explanation of what actually took place of the sort that would be provided by an historical narrative. Loose talk about an organism maximizing its fitness is somewhat disconcerting to the philosophically fastidious. Ordinarily, "fitness" means a property which, like "talent", is something that an organism itself is in no position to maximize.

More importantly from the point of view of general intellectual culture, organismal self-sacrifice is denoted by the word "altruism", giving the false impression that it has the sort of moral implications that it has when the term is used by normal human beings in ordinary discourse. Most of us would consider "reciprocal altruism" (Trivers 1971) an oxymoron: if you get something in return it is acting in self interest plain and simple. Sociobiologists have not done a very good job in their efforts to explain what they mean by "altruism" (Bertram 1982; Maynard Smith 1982; Settle 1993). Basically, however, they mean an action on the part of one organism that reduces its own fitness, while increasing that of another one. Because fitness is strictly relative, such actions can be far from edifying. Risking one's neck in the course of perpetrating an act of genocide is "altruistic" as sociobiologists define that term.

One has to be a pretty bad zoologist to confuse such "proto-ethical" behavior as a worker bee killing her brother with the real thing. The question remains, however, of whether morality proper can evolve by any known mechanism. One alternative is that some kind of enlightened self-interest might work. Another possibility is that the emergence of human society creates an environment under which virtuous persons are somehow rewarded in terms of reproductive success. Elsewhere I have suggested that a kind of unconscious artificial selection might be responsible (Ghiselin 1974b).

Be this as it may, the crass expediency that prevails in animal societies ought to deter us from attempting to derive moral imperatives from observations upon the bestial world. Research upon such matters might help us to understand our own predicament, and suggest some options, but it will never allow us legitimately to derive an "ought" from an "is". Our remote ancestors evolved in a world in which the capacity to make such distinctions had yet to come into being. The moral order is something that we have discovered, and created, and that may or may not prevail in the long run.

X. Progress

My suggestion that the moral order "may or may not prevail" underscores the point that what happens in evolutionary history is a highly contingent affair. Darwin rejected older notions about "necessary progress" but nonetheless believed that progress itself is a real phenomenon, and one that can be explained in terms of his theory (Ghiselin 1995). Progress is what one might reasonably expect, given enough time and a favorable state of the natural economy. Competition leads to diversification, with increased division of labor, both within the organism and within the natural economy. The more modified forms utilize resources more efficiently, and the archaic ones tend to be driven to extinction. Whatever features that contribute to that competitive ability tend to accumulate: and that is what is meant by "progress". It is very easy, however, to slip into outdated modes of thinking, and assume that in calling it "progress" we mean that it has to be a good thing. Counterfeiters are making progress in producing bank notes that are increasingly hard to tell from the genuine article. That is perhaps "good" for them, but not for the rest of us.

For the natural and the political economy alike, the engine of progress is technological innovation (Mokyr 1990). Indeed, cumulative technological innovation is virtually a synonym for the kind of progress that Darwin had in mind, using "technological" in a broad and metaphorical sense. If we look at the history of such optical instruments as the camera and the microscope we can enumerate any number of innovations that gave their manufacturers a competitive edge over the competition: lenses of various kinds, the iris-diaphragm, color film, etc. Comparable innovations are well documented in the evolutionary history of eyes, and students of biomechanics explain them in terms of the same kind of engineering principles. We may feel that we are stretching a term a bit when we apply the term "technology" to modes of social organization. The assembly line perhaps seems more like technology than does trial by jury. On the other hand there does exist an evolutionary theory of the common law, and one that has been modeled upon natural selection (Hirshleifer 1982). It has much in common with evolutionary approaches to the traditional areas of economics.

Progress is not a very popular idea either with sociobiologists such as Richard Dawkins (1992) or their ideological opponents, even when they freely admit that it is a real phenomenon (Gould 1988). According to Darwin's theory, the acquisition of features that give organisms a competitive edge is a highly contingent matter. An adaptive change may or may not happen, and when it does, it may take a vast amount of time. Indeed, some of the best evidence against the teleological character of evolution is the frequency of maladaptations. If we took such advocates of the "Panglossian paradigm" as Dennett (1983) seriously, we would have to reason as follows. Not only is this the best of all possible worlds, with everything in it a necessary evil, but at any given instant, it has to be better than at any other instant! The advocates of such Panglossianism are persons who view the world at a particular moment and take its "optimality" as a posit. But adaptation, the process, means change in the direction of such "optimality". If claims about adaptive significance are not to be purely hypothetical, they have to be backed up by historical data, and that means placing them in the context of an historical science.

XI. Economic History

Again, by an historical science we mean something like astronomy, or plate tectonics, with a narrative that accounts for what happens in terms of both particular events and laws of nature. Darwin's (1851, 1854) first book on evolution, the vast *Monograph on the Sub-Class Cirripedia*, was a fine example, even though the work was largely descriptive and the processual aspect was left implicit, to be only partly explicated in 1859. But we can easily read the monograph as a contribution to the *economic* history of the living world. It documents adaptive radiation and the division of labor. As a comparative anatomist, I find it very easy to conceive of reconstructing that science as a kind of economic history. It would lend new excitement to a discipline that is getting bored with drawing yet another tree-like diagram. And it would be very much like human economic history, with its stress upon the sort of innovation that gives its possessor a competitive edge.

Although such a program might appeal to a large number of evolutionary biologists, I have my doubts about economists in the traditional sense. They want to make their science look as much like physics as possible, and according to the metaphysics that is popular among physicists, science is not about history, but about laws. We biologists don't have to believe such nonsense. After all, Darwin *was* the Newton of the grass blades. Physicists don't have to believe it either, and the physicists themselves have begun to recognize the importance of the Darwinian revolution for their own intellectual lives. Economics also had the misfortune of being treated as a social science. Things would surely have been very different if it had originated as a branch of biology. But that only serves to underscore one final point: what scientists do is far more strongly conditioned by historical circumstances than most of them are aware. If there is to be an economics of scientific investigation, it too will have to be an historical science.

References

BECKER, GARY S.: *Human Capital: a Theoretical and Empirical Analysis, with Special Reference to Education*, Second Ed. 1980, Chicago (University of Chicago Press) 1975.

BERTRAM, BRIAN C. R.: "Problems with Altruism", in: R. W. WRANGHAM et al. (Eds.): *Current Problems in Sociobiology*, Cambridge (Cambridge University Press) 1982, pp. 251-267.

BOYD, ROBERT, PETER J. RICHERSON: *Culture and the Evolutionary Process*, Chicago (University of Chicago Press) 1985.

CRONIN, HELENA: *The Ant and the Peacock: Altruism and Sexual Selection from Darwin to Today*, Cambridge (Cambridge University Press) 1991.

DARWIN, CHARLES: *A Monograph on the Sub-Class Cirripedia, with Figures of all the Species. The Lepadidae; or, Pedunculated Cirripedes*, London (Ray Society) 1851.

DARWIN, CHARLES: *A Monograph on the Sub-Class Cirripedia, with Figures of all the Species. The Balanidae (or Sessile Cirripedes); the Verrucidae, Etc., Etc., Etc*, London (Ray Society) 1854.

DARWIN, CHARLES: *The Descent of Man, and Selection in Relation to Sex*, 2 vols., London (John Murray) 1871.

DAWKINS, RICHARD: *The Selfish Gene*, Oxford (Oxford University Press) 1976.

DAWKINS, RICHARD: *The Extended Phenotype: the Gene as the Unit of Selection*, Oxford (Oxford University Press) 1982.

DAWKINS, RICHARD: "Progress", in: E. F. KELLER/E. A. LLOYD (Eds.): *Keywords in Evolutionary Biology*, Cambridge (Harvard University Press) 1992, pp. 263-72.

DENNETT, DANIEL C.: "Intentional Systems in Cognitive Ethology: the 'Panglossian Paradigm' Defended", *The Behavioral and Brain Sciences*, 6 (1983), pp. 343-390.

DEWEY, JOHN: *The Influence of Darwin on Philosophy, and Other Essays in Contemporary Thought*, New York (Henry Holt & Company) 1910.

DURHAM, WILLIAM H.: *Coevolution: Genes, Culture, and Human Diversity*, Stanford (Stanford University Press) 1991.

ELDREDGE, NILES: *Unfinished Synthesis: Biological Hierarchies and Modern Evolutionary Thought*, New York (Oxford University Press) 1985.

ELDREDGE, NILES: "Information, Economics, and Evolution", *Annual Review of Ecology and Systematics*, 17 (1986), pp. 351-369.

ELDREDGE, NILES: *Macroevolutionary Dynamics: Species, Niches, and Adaptive Peaks*. New York (McGraw-Hill Publishing Company) 1989.

ELWORTHY, CHARLES: *Homo Biologicus: an Evolutionary Model for the Human Sciences*. Berlin (Duncker & Humboldt) 1993.

GHISELIN, MICHAEL T. (1974a): "A Radical Solution to the Species Problem", *Systematic Zoology*, 23 (1974), pp. 536-544.

GHISELIN, MICHAEL T. (1974b): *The Economy of Nature and the Evolution of Sex*, Berkeley (University of California Press) 1974.

GHISELIN, MICHAEL T.: "Categories, Life, and Thinking", *The Behavioral and Brain Sciences*, 4 (1981), pp. 269-313.

GHISELIN, MICHAEL T.: "Biology, Economics, and Bioeconomics", in: G. Radnitzky (Ed.): *Universal Economics: Assessing the Achievements of the Economic Approach*, New York (Paragon House) 1992, pp. 71-118.

GHISELIN, MICHAEL T.: "Darwin, Progress, and Economic Principles", *Evolution*, 49 (1995), pp. 1029-1037.

GHISELIN, MICHAEL T.: *Metaphysics and the Origin of Species*, Albany (State University of New York Press) 1997.

GOULD, STEPHEN JAY: "On replacing the idea of progress with an operational notion of directionality", in: M. H. NITECKI (Ed.): *Evolutionary Progress*, Chicago (University of Chicago Press) 1988, pp. 319-338.

HERTWIG, OSCAR: *The Becoming of Organisms: toward the Refutation of Darwin's Chance-Theory through that of Law in Development*, 3rd ed., Jena (Verlag von Gustav Fischer) 1922. [In German.]

HIRSHLEIFER, JACK: "Natural Economy Versus Political Economy", *Journal of Social and Biological Structures*, 1 (1978), pp. 319-337.

HIRSHLEIFER, JACK: "Evolutionary Models in Economics and Law: Cooperation Versus Conflict Strategies", *Research in Law and Economics*, 4 (1982), pp. 1-60.

MAYNARD SMITH, JOHN: "The Evolution of Social Behaviour - a Classification of Models" in: R. W. WRANGHAM et al. (Eds.): *Current Problems in Sociobiology*, Cambridge (Cambridge University Press) 1982, pp. 29-44.

MICHOD, RICHARD E./BRUCE R. LEVIN (Eds.): *The Evolution of Sex: an Examination of Current Ideas*, Sunderland (Sinauer Associates) 1988.

MILNE EDWARDS, H.: *Introduction to General Zoology, or Considerations upon the Tendencies of Nature*, Paris (Victor Masson) 1851. [In French.]

MOKYR, JOEL: *The Lever of Riches: Technological Creativity and Economic Progress*, New York (Oxford University Press) 1990.

RADNITZKY, GERARD (Ed.): *Universal Economics: Assessing the Achievements of the Economic Approach*, New York (Paragon House) 1992.

SETTLE, TOM: "'Fitness' and 'Altruism': Traps for the Unwary, Bystander and Biologist Alike", *Biology and Philosophy*, 8 (1993), pp. 61-83.

TRIVERS, ROBERT L: "The Evolution of Reciprocal Altruism", *Quarterly Review of Biology*, 46 (1971), pp. 35-57.

Chapter 2

Coercion

TIM CLUTTON-BROCK and GEOFF A. PARKER

Introduction

Over the last three decades, research in behavioural ecology has focussed on the choices that individuals make over food (Stephens/Krebs 1986), mating opportunities (Andersson 1995), parental care (Clutton-Brock 1991) and cooperation with other group members (Brown 1987; Clutton-Brock/Parker 1995b). Yet, in social species, most individuals rarely have the opportunity to choose freely because their options are constrained by the behaviour of dominant animals (Clutton-Brock/Parker 1995b). Coercion is common in many contexts where conflicts of interest occur, but is particularly important in interactions between the sexes, where it often has substantial costs to females. In a recent paper (1993), Smuts and Smuts argue that sexual coercion should be regarded as a third form of sexual selection, separate from intrasexual competition and intersexual mate choice. Whether or not this is accepted, it is clear that sexual coercion has not received the attention that it deserves.

Dominant individuals can constrain the options of other group members either by manipulating their physical or social environment (for example, by driving potential mates, allies or recipients of investment out of the groups) or by manipulating their behaviour directly (Smuts/Smuts 1993; Clutton-Brock/Parker 1995a). In this review, we examine three behavioural tactics (use of direct force; harassment; and punishment) used by males to manipulate the behaviour of females, outlining areas of theory relevant to their evolution. Females commonly respond with behaviour that minimises both the costs of male tactics and the ability of males to constrain their behaviour (Smuts/Smuts 1993) and their tactics can have an important influence on the evolution of male behaviour.

I. Direct Force

Stronger or more dominant animals frequently constrain the decisions of weaker or more subordinate individuals by the direct application of force. In a variety of animal species, individual males pursue females and, if they can catch them, restrain them and force them to copulate (Smuts/Smuts 1993). For example, in wild orangutans *Pongo pygmaeus* most copulations by subadult males and almost half of all copulations by adult males occur after fierce female resistance has been violently overcome by the male (Mitani 1985). Similar cases of forced copulation have been observed in many other polygynous vertebrates lacking long-term bonds between the sexes (McKinney et al. 1983; Emlen/Wrege 1986) as well as in invertebrates (Thornhill 1980; Arnqvist 1989). By contrast, forced copulation appears to be rare in monogamous species or polygynous species where a single male can control mating access to one or more females effectively (Smuts/Smuts 1993).

Where males can force females to copulate with them, conflicts of interest between the sexes are likely to occur (Trivers 1972; Parker 1979). A female may gain an increment to her fitness by delaying or refusing mating if this allows her to compare potential mating partners or to mate with a superior male. Alternatively, there may be advantages to her in delaying mating until a higher proportion of females are ready to conceive. A male is likely to gain a larger increment by forcing the female to mate with him. Where there is a risk of other males locating the female, there may be strong selection for males to enforce acquiesence on females as rapidly as possible.

Sexual conflict of this kind has the potential to lead to the evolution of sexual dimorphism, fuelling arms races that can generate substantial costs to both sexes (see Box 1). For example, selection for the ability to force females to mate may help to explain the evolution of large male body size in males (Smuts/Smuts 1993). Conversely, selection on females to avoid male harassment and to maintain their ability to exercise mate choice may sometimes favour increased female size.

Box 1: Logic of a simple arms race game between the sexes

Suppose that when a male meets a female, it pays the male to mate but not the female because she may already contain sperm or prefers an alternative mate. etc.). The value, V, of winning the contest (i.e. mating for the male; not mating for the female) differs for the two opponents, so that $V_m \neq V_f$. Suppose that the male wins if his arms level, R, is greater than that of the female, i.e. if $R_m > R_f$ while the female wins if her arms level exceeds that of the male. Over evolutionary time, selection can operate on a continuous range of expenditure on armament in each sex. If a female has an arms level that costs J, then to equal this level, a male must pay an armament cost K. The relative costs of armament are also expected to be unequal, so that $J \neq K$. If a strategy (a choice of expenditure, J or K) prescribes an arms level exactly, there is no evolutionarily stable strategy ESS for this game (Parker 1979; Maynard Smith 1982) but an ESS can occur if there is sufficient random or environmental variation in the arms level attained by a given strategy (Maynard Smith 1982). For example, suppose that if a male expends K, its realised arms level, $R(K)$, is not prescribed exactly, but has the probability distribution $p_m(R)$, and similarly $p_f(R)$ for the female. The ESS depends on the forms of the distributions $p_m(R)$ and $p_f(R)$: it is predicted that at the ESS, males spend K^* and females J^*. These generate overlapping ESS distributions of realised armaments in the male and female populations: $p_m{}^*(R)$ and $p_f{}^*(R)$. A hypothetical example in which males generally win is shown in Fig. 1 (Clutton-Brock/Parker 1995b). A female with arms level $R_f < u$ always loses, and a male with arms level $R_m < v$ always wins. The outcome of the contest in the range u, v is ambiguous and depends on the arms level of each opponent.

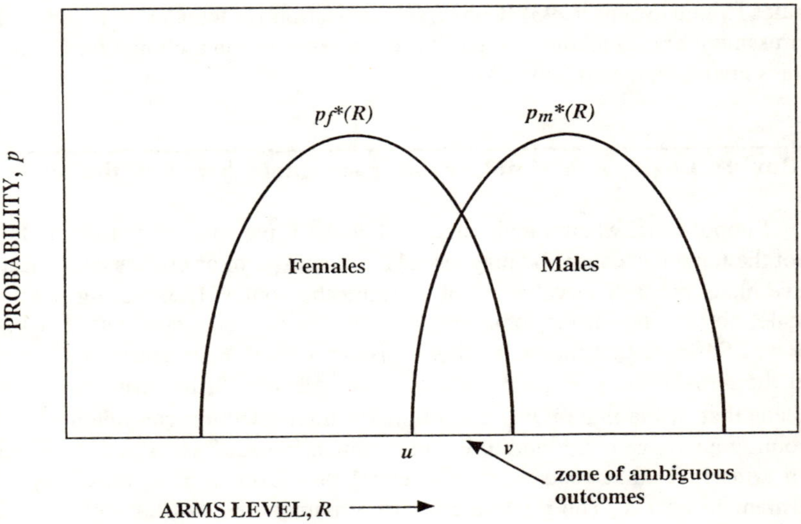

It is difficult to predict which sex will occupy the winning role for this depends on asymmetry in the values of winning, V_m, V_f, and in the costs of increasing armaments. In general, the male stands to gain *n* offspring by mating, whereas the female stands only to gain a change in the quality of the *n* offspring by not mating, or to gain from avoiding some cost of mating (time, energy, or risk). Hence it is likely that $V_m > V_f$. However, the relative costs of armament may often lie in favour of the female. For example, in many species, it may be more expensive for a male to achieve a forced copulation than for the female to prevent it. Suppose that armament is exactly balanced, so that the male's and female's chances of winning are equal. If the female increases her armament by one small unit, costing C_f, we need to know the cost, C_m, to the male of restoring the balance. When $C_m > C_f$, this tends to offset the effect of $V_m > V_f$, making predictions about the winning role difficult. Sex *i* is likely to be in the winning role if $V_i / C_i > V_j / C_j$ (see also Box 2). An additional complication is that intra-sexual selection can drive male body size above that of female body size; leading to a situation where the costs of increasing armaments to males are subsidised by other benefits.

II. Harassment

In many animal species, individual males that locate potentially receptive females persistently court and attempt to mate until the female either leaves or copulates with them (Thornhill/Alcock 1983). When males increase their probability of mating (and, hence, their fitness) by harassing females, while females can benefit by withholding matings, conflicts may usually approximate to asymmetric wars of attrition (Parker 1979; Hammerstein/Parker 1982), where two contestants compete by some form of persistence.

Theoretical models of asymmetric wars of attrition (Clutton-Brock/Parker 1995b) predict that, where males can increase their mating success and fitness by harassing females, harassing tactics evolve to a point where they have substantial costs to both sexes (see Box 2). Costs to females may be particularly high if the Operational Sex Ratio is strongly male-biased and multiple males court females simultaneously, each male persisting until the costs of continuing outweigh the benefits so that the cumulative costs for the female are very high. Moreover, where a male that locates a receptive female is likely to lose a mating if she does not 'agree' to copulate immediately, males may be selected to raise the costs of continued refusal by attacking them until they cooperate (Smuts/Smuts 1993).

Empirical studies provide abundant evidence of the costs of male harassment to females (Smuts and Smuts 1993; Clutton-Brock and Parker 1995b). Sexual harassment commonly reduces female feeding activity (Smuts/Smuts 1993). For example, in feral horses, females living in non-territorial harems (where harassment is relatively common) are more frequently disturbed while feeding and show lower breeding success than females associated with territorial males that can provide effective protection against other males (Rubenstein 1986). Harassment can sometimes have lethal consequences for females: for example, female ducks are sometimes drowned by competing males (McKinney et al. 1983), while female ungulates may be killed in fights between multiple males (Clutton-Brock et al. 1993). Some of the clearest evidence of the costs of harassment comes from research on captive insects. Experiments with *Drosophila* show that exposure to males has costs to female survival, arising partly from repeated courtship and partly from repeated mating (Fowler/Partridge 1989; Chapman et al. 1993).

When the risk of dangerous harassment is high, female behaviour is commonly adapted to evading harassment or minimising its costs (Rubenstein 1986; Smuts/Smuts 1993). For example, in polygynous ungulates that live in large, mixed-sex herds where there is no stable dominance hierarchy among males, oestrus females commonly leave the herd and seek refuge in mating territories defended by males (Rubenstein 1986; Clutton-Brock et al. 1993). Where males defend consort partners rather than territories, females may also prefer to associate with males that can provide effective protection against conspecifics (Wrangham 1980; Smuts 1986a). For example, female savanna baboons commonly form special relationships with particular males that provide protection to them or their offspring, associating preferentially with them throughout the year and mating with them when they enter oestrus. In some damselflies and butterflies, the risk of male harassment has even led to the evolution of cryptic female morphs that mimic males, reducing frequency of harassment (Cook et al. 1994; Robertson 1985).

Like forced copulation, harassment by males may fuel arms races between the sexes. In social primates, mature females sometimes form coalitions to protect each other against males (de Waal 1982; Smuts/Smuts 1993). Females will gang up on males that attack other females or attempt to force them to copulate or threaten their families (de Waal 1982). These coalitions form rapidly and - in contrast to coalitions formed to compete for resources - commonly involve unrelated females. It is clear that they can be effective, in some cases preventing immigrant males from joining their group or even forcing residents to emigrate (Packer/Pusey 1979; Cheney 1983a; Cheney 1983b; Mesnick, in press).

Box 2: Game theory models applicable to sexual harassment

Box 1 argues that sexual arms races are likely to proceed to a point where both sexes make substantial expenditure on armament, and where the outcome of interactions is likely to be biased in favour of one sex but will not be entirely predictable since the balance of power over the range u, v, depends on the individuals involved. There are essentially two evolutionary game theory models that can be applied to sexual harrassment: the asymmetric war of attrition (Hammerstein/Parker 1982) and the sequential assessment game (Maynard Smith/Parker 1976). The details of how these two approaches can be applied to sexual harrassment are given in Reference 5. Both models examine how long individuals should persist in a contest: the main constraints being the cost of persistence and the value of winning.

Both models assume that information about the power asymmetry is imperfect.

At the start of an asymmetric war of attrition, each contestant estimates (i) the rate at which relative costs will be accrued during the interaction (c_A, c_B), and (ii) the relative values of winning (V_A , V_B), and chooses a persistence time t (time to continue harassment for males, or a time to continue with resistance for females). The winner is the individual which has chosen the higher persisitence time. The ESS for the asymmetric war of attrition is shown in Fig. 2. It consists of a distribution of persistence time choices which is dependent on the role that an opponent estimates itself to occupy in the contest. Where *A* is the 'winning role' and *B* is the 'losing role'. If a contestant believes itself to be in role *A*, it chooses a persistence time from the probability distribution $p_A(t)$, and if it believes it self to be in role *B*, it chooses from the probability distribution $p_B(t)$. These two distributions do not overlap, but there is no gap between them; they are separated at the separation time s . An individual perceiving itself to be in the winning role *A* always chooses a persistence time greater than that of an individual perceiving itself to be in role *B* , so that an individual choosing from $p_A(t)$ always beats one choosing from $p_B(t)$. Roles *A* and *B* are defined by the following relationship:$V_A / c_A > V_B / c_B$. *B* has the lower "fitness budget for fighting": if both contestants persist, *B* would be first to spend more in contest costs than it could gain by winning.

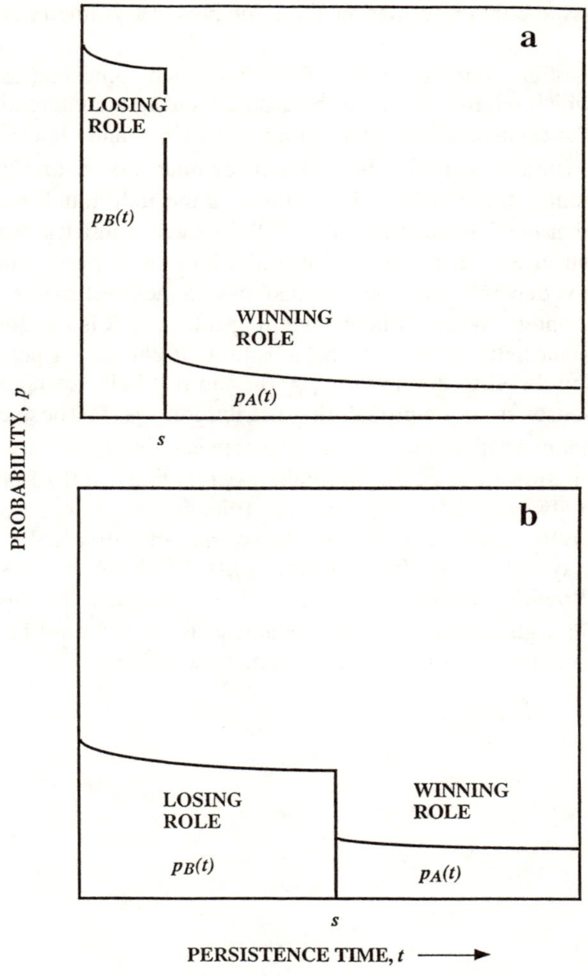

If both contestants have perfect information about their roles, there is no ESS. But perfect information about the values of V or c seems highly unlikely. Provided that errors can be made about roles, so that sometimes both contestants assess themselves to be in the same role, the ESS is given by the two distributions $p_A(t)$, $p_B(t)$. The greater the probability of error, the higher the separation point s, and hence (i) the greater the expected contest duration, and (ii) the greater the variance in outcome (wins by the contestant actually in role B become more frequent). Errors will be commonest where opponents are equally matched, leading to higher contest costs. If errors are extremely rare, the opponent in the losing role B will give up almost immediately (separation point s is close to zero), and will be beaten almost invariably by the opponent in A. Since, usually, the value of winning will generally be greater for males ($V_m > V_f$), but it may often be less expensive for a female to prevent mating than for a male to achieve it against female resistance ($c_m > c_f$) it is not easy to predict which sex will occupy the winning role unless the male is much larger than the female.

In the sequential assessment game, opponents continue to gain information about relative fighting abilities (and hence relative contest costs, c_A , c_B) during the contest itself, which consists of a sequence of successive steps. At each step, an individual's estimate becomes more accurate as the size of the sample on which the estimate is based increases. The ESS is a threshold value for the estimate of relative contest costs at which the decision to continue gives the same expected payoff as the decision to give up. It is derived numerically, using a form of dynamic optimization, and can be represented as a pair of switching lines (one for each opponent) which increase with the number of steps. A hypothetical switch line is shown in Fig. 3 (Clutton-Brock/Parker 1995b). The contest ceases when the first opponent crosses over its switching line and quits: this is likely to be the weaker opponent, but note that (as in the asymmetric war of attrition) the opponent that is in reality stronger may (by chance) be the one to cross its switching line first. High uncertainty causes contestants to be unwilling to give up early; the cost of continued sampling is paid in order to get a more accurate estimate. If the costs per step are low relative to the value of winning (V), long contests are advantageous in order to gain more accurate estimates, and if the true fighting abilities are similar, contests continue longer than if there is a large difference. The switch line is lowered as V increases - the male is likely to have the higher V (see above), but because contest costs for the male may be high, too, this will tend to raise the male switch line. Thus this model generates conclusions that are virtually identical to that for the asymmetric war of attrition. Sexual conflicts obeying the asymmetric seqential assessment game can be protracted and expensive to both sexes if the arms levels overlap sufficiently (Box 1).

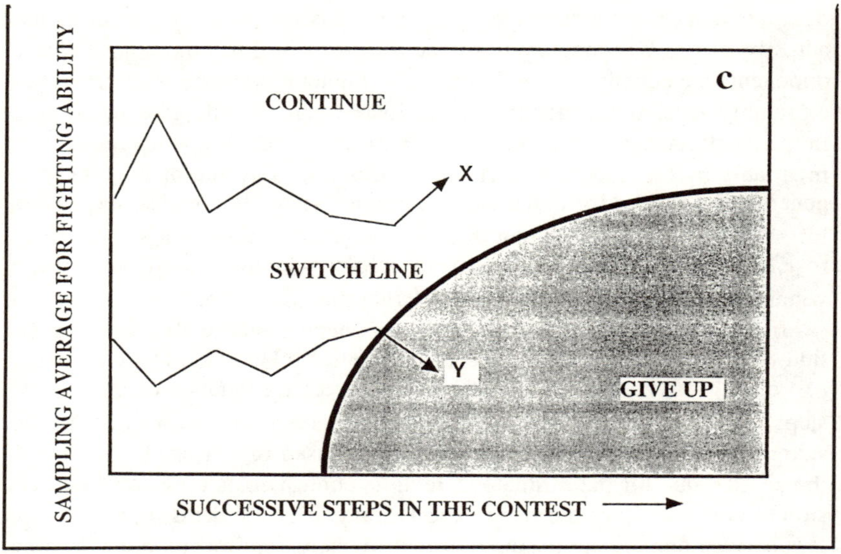

III. Punishment and Intimidation

Punishment involves the retaliatory infliction of a fitness reduction which benefits the punisher because victims learn to avoid repeating damaging behaviour in the future, or in extreme cases, because they are eliminated or forced to emigrate (Clutton-Brock/Parker 1995a). The simplest case of punishment is where a subordinate transgresses against a dominant, receives a punishment and subsequently cooperates, to the benefit of the dominant's fitness (Box 3). Unlike harassment, the benefits of punishment (to the punishers) are delayed and, consequently, punishment is likely to evolve only where the same individuals interact repeatedly.

Box 3: Interactions between non-kin

Figure 4 shows six categories of interactions between non-kin. A minus sign represents a loss of fitness, a plus sign a fitness gain. Arrows indicate the original initiator and recipient. In punishing interactions, a selfish action by the first player is followed by the first player being punished by the second. Subsequently, the first player modifies his/her behaviour so that it increases the fitness of the punisher at some cost to itself. Punishing tactics may also be used to coerce or augment altruistic behaviour in subordinates (see text). All six categories of interaction also occur between kin, though indirect benefits may alter the magnitutde of fitness gains and losses.

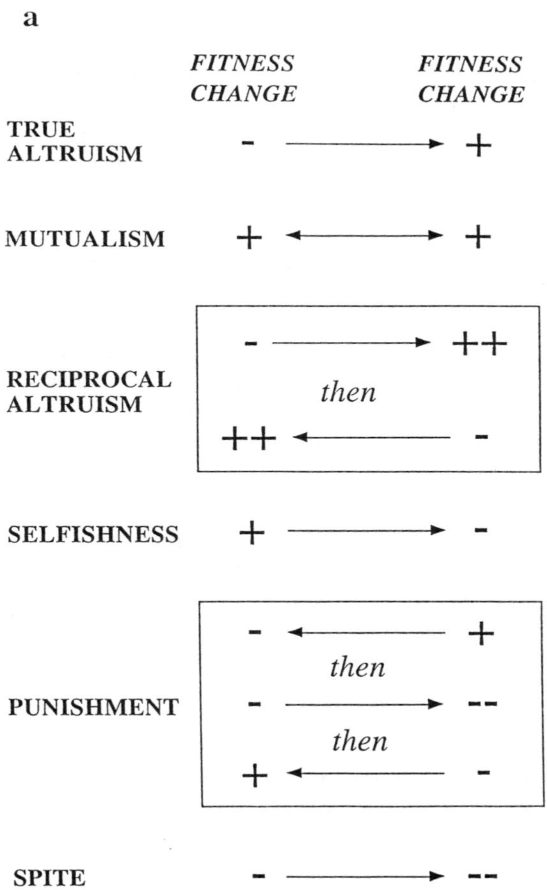

Game theory models confirm that punishing tactics will often be evolutionarily stable. Box 4 considers situations where two individuals differ in fighting ability (so that one can punish the other without fear of retaliation), while Box 5 considers the more complex situation where individuals can retaliate to punishsment, incorporating retaliation as a variant of the simple Hawks-Doves game with asymmetric animal contests (Maynard Smith 1974; Maynard Smith/Parker 1976; Maynard Smith 1982; Hammerstein 1981; see Box 2). Both models initially focus on situations where a subordinate can choose whether to transgress(T) against a dominant or to desist(D) from doing so while the dominant can choose whether to punish them or not to do so(NP). Subsequently, we consider cases where roles are reversed and subordinates can choose whether or not to punish the transgressions of dominants. In both models, 'commonsense' ESS's, where dominant individuals punish subordinates who subsequently avoid transgressing, and where dominants transgress against subordinates without fear of punishment or retaliation, are most likely to develop. In the context of sexual interactions, transgress(T) represents a refusal to mate while desist(D) represents an agreement to do so.

Punishing tactics are widespread (Clutton-Brock/Parker 1995a). In many social mammals, males punish females that refuse to associate with them with various forms of physical violence (Nadler 1982; Nadler/Miller 1982; Nadler 1988; Smuts/Smuts 1993; Clutton-Brock/Parker 1995a). For example, Goodall (1986) describes how male chimpanzees will repeatedly attack females in the early stages of consort formation until they become more cooperative and follow the male closely. Males will also punish females that refuse their mating attempts. Dominant male primates may also punish females for approaching, courting or mating with subordinates or with males from other groups (de Waal 1982; Lindburg 1983; Goodall 1986; Manson 1994), while similar behaviour has been observed in some birds (Westneat et al. 1990). There is widespread anecdotal evidence that male aggression can have substantial costs to females (Smuts/Smuts 1993).

Though firm experimental evidence is lacking, several studies suggest that aggression directed at receptive or near-receptive females may increase the chance that they will mate with aggressors, though it does not always do so (de Waal 1982; Nadler/Miller 1982; Smuts/Smuts 1993). Male Japanese macaques (*Macaca fuscata*) that show aggression to their consorts during the mating season were significantly more likely to mate with them than males that did not show agression towards their consorts (Enomoto

1981). Detailed studies of captive lowland gorillas show that females learn to present (for copulation) more frequently to more aggressive males, thereby reducing the frequency with which they were attacked (Nadler/Miller 1982). The high rates of presenting to aggressive males did not arise because females were attracted to more aggressive males, for they consistently avoided them.

Box 4: Punishment with extreme asymmetry in dominance

Imagine a game consisting of N separate interactions (rounds) between two animals with a pronounced difference in fighting ability where the dominant can punish the subordinate without significant risk of retaliation. Fig. 1 outlines the payoff matrix. The cumulative payoffs over the N rounds to the dominant (Player 1) are shown in top right of each square, payoffs to the subordinate (Player 2) are shown on the lower left. On each round, the subordinate can take an action which increases its own fitness at the dominant's expense. It may either Transgress (T) by committing the action, or may Desist (D), favouring the dominant's interests (in the context of sexual interactions, T translates to a refusal to mate, D to agreeing to copulate). If the subordinate plays D, it loses $-c_1$ fitness units and the dominant gains $+g_2$. The dominant may Punish (P) the subordinate if it plays T, or may Never Punish (NP) whatever the subordinate does. The Dominant never plays Punishment if the subordinate Desists. Punishing has a small cost of $-d_1$ units to the dominant while being punishedcosts the subordinate $-d_2$. The subordinate must learn to Desist, and learning takes j rounds, so that a D-player first transgress for j rounds than plays Desist for $k = N - j$ rounds if it is first punished. Thus against a P-player, the payoff of a D-player $+jg_2 - jd_2$. Against an NP-player, a D-player continues to Transgress and never switches to D, and hence gains $+Ng_2$, whilst the NP player loses $-Nc_2$. Payoffs are therefore the same as with T and NP. Punishment occurs on each round with T and P, so T's payoff is $+Ng_2$ $-Nd_2$ and P loses $-Nd_1$ $-Nc_1$.

DOMINANT (Player 1)

SUBORDINATE (Player 2)	**P**	**NP**
D	Player 1: $-jd_1$ $-jc_1$; Player 2: $-jd_2$ $+jg_2$	Player 1: $-Nc_1$; Player 2: $+Ng_2$
T	Player 1: $-Nd_1$ $-Nc_1$; Player 2: $-Nd_2$ $+Ng_2$	Player 1: $-Nc_1$; Player 2: $+Ng_2$

The punishment strategy "Desist when 2, Punish when 1" is an ESS provided that (i) the summed benefits exceed the total punishment effort ($kc_1 > jd_1$) to prevent player 1 from playing NP, and that (ii) each punishment costs more than the benefit of transgressing ($d_2 > g_2$) to prevent player 2 from playing T. But the non-punishment strategy "Transgress when 2, Never Punish when 1" also has some stability, since (i) player 1 is prevented from playing P (since $0 > -Nd_1$) and (ii) any tendencies to show D unconditionally (i.e. against NP) may prevent D spreading by drift, and so maintain NC. Thus the strategy (D, P), and (with additional assumptions) (T, NP), may both be ESS's. The other possibilities (D, NP) and (T,P) cannot be ESS's since it cannot pay to desist if there is no chance of punishment, or to punish if there is no chance of receiving benefit.

Note that in this simple example, player 2 can either act selfishly and Transgress against the interests of player 1, or Desist from acting selfishly so as to favour the interests of player 1. We could have called the choices of player 2 "Non-Cooperate" or "Cooperate" to describe the case where the alternatives involve positive action to favour player 1, rather than positive action to favour self (player 2). The conclusions are exactly analogous (Clutton-Brock/Parker 1995a): cooperation maintained by punishment (C,P) or non-cooperation with non-punishment (NC,NP) can be ESSs.

Box 5: Weaker asymmetries: ESSs with punishment and retaliation

In the simple game in Box 4, player 1 is cast in the role of dominant and player 2 as subordinate, and it is assumed that because of this asymmetry, cooperation can flow only from the subordinate to the dominant. Suppose that the difference between the two individuals is weaker, so that either may retaliate against punishment - what types of ESS might be possible?

Clutton-Brock and Parker (1995a) considered the game as a sequence of successive steps or moves between players 1 and 2 (Fig. 6). Steps 1 and 2 occur in each round, as in the simple game (Box 4) - at step 1, player 2 has the opportunity to Transgress (T), or to learn to Desist (D) towards player 1, but j training rounds of punishment are required before player 2 switches from T to D for the remaining k rounds ($j + k = N$). At step 2, player 1 may Punish (P) if 2 played T, or Never-Punish (NP).

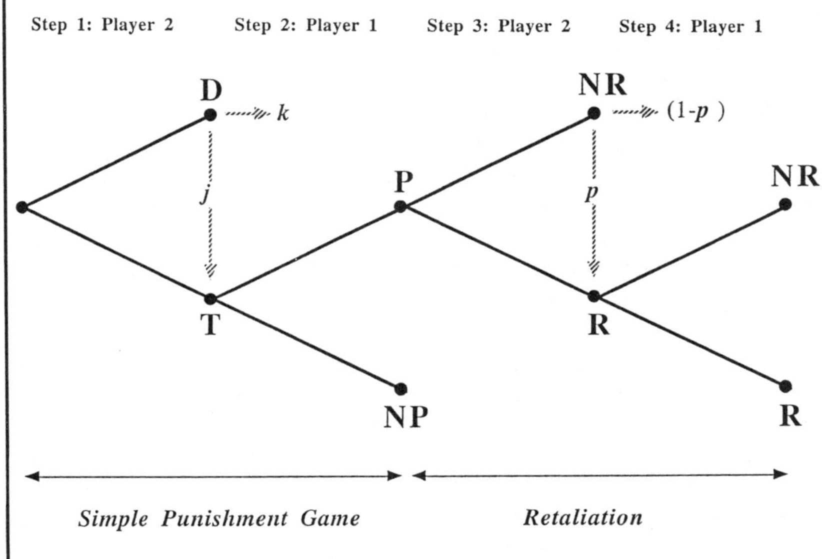

Step 3 occurs in a round only if player 1 Punishes at step 2: player 2 may then Retaliate (R) by making an escalated attack on player 1, or may Non-Retaliate (NR). If he chooses NR, he nevertheless escalates (i.e. plays

R) with a low probability, $p \to 0$. p may relate to testing relative fighting ability, or simply to mistakes about roles (Stephens and Krebs 1986; Clutton-Brock and Parker 1995a). If player 2 does not retaliate at step 3, the game stops and payoffs remain unaltered. But if Retaliate is chosen, then (step 4) player 1 may also choose to Retaliate, or to Non-Retaliate (but with probablity 1.0). In steps 3 and 4, R always beats NR without escalation or injury, whatever the relative fighting abilities of the two contestants. If both choose R, an escalated contest ensues and results in an injury to one of the players, after which the escalated contest ceases. The probability that a player is injured, and hence loses, depends on relative fighting abilities. Steps 3 and 4 resemble the asymmetric "Hawks/Doves" game (Hammerstein 1981).

In a punishment ESS, player 1 will be giving punishment and later receiving altruism. If, either after fighting or by giving up, player 1 loses when player 2 retaliates against punishment, there will be social consequences: player 1 will have lost status and player 2 will have gained status. Clutton-Brock and Parker (1995a) incorporated this effect into the payoffs by assuming that if player 1 loses, for a small number of rounds after losing, player 2 then transgresses and player 1 plays Not Punish. If player 1 wins, there are no additional payoffs, and the next round proceeds normally.

Clutton-Brock and Parker (1995a) examined how relative fighting ability ("dominance") of the two players influences ESS conditions. There are four candidate ESSs that involve punishment. These can be listed in terms of players' tactics at successive steps: e.g. P/R, D/NR is "if player 1, Punish at step 2, Retaliate at step 4; if player 2, Desist at step 1, Non-Retaliate at step 3". An ESS in which neither player retaliates is impossible, and that one in which both retaliates is possible only if injury is trivial so that both players stand to gain by escalating (see also Hammerstein 1981). If injury is sufficiently expensive, only ESSs in which one opponent is prepared to retaliate can be stable, and far the most likely ESS is to play P/R when player 1, and D/NR when 2 . This has formal requirements that are very similar to the simple game in Box 4: (i.e. that $kc_1 > jd_1$; $d_2 > g_2$), and also requires that player 1 has a positive expectation from escalation, and that player 2 has a negative expectation: ie. it is likely to involve punishment by a strong individual of a weak individual ("commonsense", sensu Maynard Smith and Parker (1976). The alternative in which a weaker player 1 punishes a stronger player 2 ("paradoxical") can exist, but is less likely to be stable.

What about ESSs for non-punishment? Once again, the ESS is likely to be affected by fighting ability and punish and transgress cannot coexist together as an ESS (Box 4). Clutton-Brock and Parker (1995a) examined two forms of transgression when retaliation is possible: Weak Transgression (WT) is an opportunistic attempt to transgress against player 1, which is successful only if player 1 replies with Non-Retaliate (NR). If player 1 replies with Retaliate (R), WT gives up immediately without gain or loss to either opponent. In contrast, Strong Transgression persists if player 1 plays Retaliate, until one of the players is injured. As before, players differ in fighting ability and this determines the probability of injury.

Player 1

Player 2	**R**	**NR**
WT	Player 1: 0; Player 2: 0	Player 1: $-Nc_1$; Player 2: $+Ng_2$
ST	Player 1: $-NxD_1$; Player 2: $+Nxg_2$ $-N(1-x)D_2$	Player 1: $-Nc_1$; Player 2: $+Ng_2$

There are again four candidate ESSs. Of these WT,NR can be discounted immediately, and ST,R requires that injury is trivial. The remaining two ESSs both obey a commonsense format: WT,R is stable if player 1 has a positive expectation from escalation (favoured if player 1 is stronger), and ST,NR is stable if player 2 has a positive expectation (favoured if player 2 is stronger). Thus although the possibility exists for a weak player to show ST or R against a strong player, it is much more likely that the dominant will keep his resource (WT,R), and take resources from the subordinate (ST,NR), depending on which role he is in.

The game is characterised by two asymmetries: (i) the fighting ability difference ("dominant"; "subordinate") between player 1 and 2; and (ii) the arbitrarily-assigned role difference that gives one player the opportunity to Transgress and the other to Punish. Depending on circumstances, player 1 may have the opportunity to Transgress against player 2, or player 2 the opportunity to transgress against 1. If we relax our assumption that altruism can flow only from subordinate to dominant (see Box 4) then, depending on the circumstances, roles can be reversed. The reversed roles must be considered as a separate game. Now it becomes possible for one state to apply when the dominant is in the role of potential punisher, and another when the subordinate is the potential punisher. Possible ESSs are shown in Fig. 2 (after Clutton-Brock/Parker 1995a) as vertical pairs of strategies (one for dominant, one for subordinate). The second vertical pair of each cluster gives the possible ESS when roles are reversed. The shaded/unshaded rectangles indicate roles: shaded = potential cooperator, and unshaded = potential punisher. Thus (at least theoretically) it is possible (1) for both players to show Desist-Punish strategies reciprocally ("mutual punishment") as the opportunities occur, or (2) for both players to show Selfish-Never Punish strategies ("mutual non-punishment", 2). Alternatively, "commonsense punishment" (3) occurs if only the dominant individual receives cooperation by punishing, but when the dominant has the opportunity to show altruism towards the subordinate, he plays selfishly without risk of reprisal. "Paradoxical punishment" (4) may exist in which the flow of altruism is from the subordinate to the dominant; selfishness (without risk of reprisal) is meeted out by the subordinate to the dominant.

Biologically, states 1 and 4 would be highly unusual: punishment is usually conditional upon role, as in 3, or non-existent, as in 2. Reasons why this should occur are evident from the retaliation model. Mutual Punishment (1) is rather unlikely since it requires flow of punishment in both directions (from the subordinate to the dominant, as well as vice versa), but cannot be discounted if opponents have rather similar fighting ability. Mutual Non-Punishment (2) is feasible, though the dominant player is likely to maintain his resources by force, and the subordinate is unlikely to give up his resources without retaliation. A form of Commonsense Punishment (3) in which the dominant punishes the weaker subordinate, and takes resources by force (and without risk of retaliation from the subordinate) when roles are reversed, is also feasible. Paradoxical Punishment (4) is unlikely, but cannot be discounted. The range of possibilities is actually richer than that suggested by Fig. 2. Less likely states include "Commonsense Punishment, Paradoxical Transgression", i.e. the dominant punishes a subordinate, but when roles are reversed, is not prepared to transgress strongly against the subordinate. Similarly there may be "Paradoxical Punishment, Commonsense Transgression", or "Paradoxical Punishment, Paradoxical Transgression", etc. But the most likely states are a form of Mutual Non-Punishment in which the dominant takes resources from the subordinate but not vice-versa, or a form of Commonsense Punishment in which the dominant punishes the subordinate, yet transgresses strongly against it without reprisal when roles are reversed.

IV. Punishment in Other Contexts

Dominant individuals use punishing tactics to control the behaviour of subordinates in a wide variety of social contexts where conflicts of interest occur. For example, during the establishment of dyadic dominance relationships, two individuals frequently trade attacks or threats until one of them establishes a superior position. This is often followed by a period of consolidation when most attacks or threats are directed by the dominant at the subordinate. Subsequently, subordinates usually avoid contests, and mild threats by dominant individuals are sufficient to constrain their behaviour. However, from time to time, subordinates test dominants, probably because this allows them to check for changes in the dominant's fighting ability, which would permit them to reverse the relationship. Unsuccessful challenges commonly elicit attacks by the dominant, followed by a period of heightened sensitivity to any failure by subordinates to respond to mild threats (Rowell 1972; de Waal 1982; Datta 1983; Mendoza 1993).

Punishment of offspring by parents is common in many social vertebrates. For example, female mammals will bite, kick or slap juveniles that persistently attempt to suck or solicit other forms of investment (Clutton-Brock et al. 1982; Clutton-Brock 1991). Punishment can have considerable costs to offspring: in moorhens and coots, parents usually feed whichever offspring is closest to them but discourage 'greedy' young that persistently follow them closely after they have fed by picking them up and shaking them (Horsfall 1984; Leonard et al. 1988). In some cases, persistent young are even killed by their parents (Horsfall 1984).

Several recent studies indicate that punishing tactics are used to discourage defection in cooperative breeders and eusocial species. In some primitively eusocial wasps, dominant workers may inherit the queen's role, and conflicts of interest consequently arise between queens and dominant workers over how much time and energy the latter should invest in rearing sibs (Strassman/ Mayer 1983). Queens are regularly aggressive towards inactive workers, chasing, biting, grappling or bumping them, and experimental removal or cooling of queens rapidly lowers the level of worker activity (Reeve/Gamboa 1983; Reeve/Gamboa 1987). Similar behaviour occurs in the primitively eusocial naked mole-rat where queens direct repeated aggression at lazy workers (Reeve 1992). And in superb fairy wrens (*Malurus cyaneus*), helpers that are experimentally removed from the group at times when the group is raising young (when they play a major role in providing food) are usually attacked and harassed by the dominant male on their return, while helpers removed during the non-breeding season are never attacked (Mulder/ Langmore 1993).

In social primates, aggressive exchanges often involve kin of the principal protagonists. In vervet monkeys (*Cercopithecus aethiops*), adult females who have been displaced from food sources may seek out and attack their displacer's offspring (Cheney/Seyfarth 1986; Cheney/Seyfarth 1990). In macaques, members of different matrilineal groups ally with each other and individuals that have been displaced or attacked by members of another matriline commonly respond by attacking a vulnerable member of their aggressor's matriline (Cords 1991; Aureli et al. 1992). Attacks on a member of one matriline are commonly followed by retaliation against members of the aggressor's matriline by relatives of the victim (Aureli et al. 1992). 'Redirected' aggression of this kind may also involve unrelated allies. In savannah baboons (*Papio cynocephalus*), males commonly develop close 'friendships' with particular females and individuals that have been involved in fights may seek out and threaten or chase their attacker's 'friend' (Smuts 1985; Smuts/ Smuts 1993). Vice versa, males may threaten or attack other males or females that attack or harass their 'friends (de Waal 1982; Smuts 1985).

Punishing tactics are often associated with behaviour that appears to minimise the risks of repetition or retaliation. Where relationships are asymmetrical and there is a danger that a dominant animal will repeat an attack, selection may favour rapid reconciliation by subordinates following an attack. Ritualised reconciliative behaviour, involving affiliative behaviour or close physical contact shortly after an aggressive interaction, occurs in many social mammals while other species lack ritualised gestures but still show a high probability of affiliative interactions following an aggressive interaction (de Waal 1993). Recent studies suggest that reconciliative behaviour reduces the probability of further conflict. After reconciliative behaviour has occurred, chimpanzees are less likely to resume conflicts while, in long-tailed macaques (*Macaca fascicularis*), victims are less likely to threaten or attack members of their aggressor's matriline (de Waal 1993). Experiments with pairs of long-tailed macaques show that pairs which were given the opportunity to show reconciliative behaviour after induced conflicts subsequently showed closer proximity to each other than pairs that had been experimentally prevented from reconciling (Cords 1991). Reconciliative behaviour may also involve contestants' kin or allies: following conflicts, both contestants may seek reconciliation with members of each others' kin groups, while their relatives may also seek reconciliation with each other (de Waal 1989, 1993).

Abstract

Though models of sexual selection commonly assume that females can choose freely between potential mating partners, males often constrain the reproductive options of females by the direct use of force, by harassing them until they cooperate or by punishing behaviour likely to damage their interests while females commonly adopt tactics that allow them to evade male control. Tactics for exerting (and evading) control can be expected to a point where they have substantial costs to both sexes. Punishing tactics are not confined to interactions between the sexes and are used to constrain the demands of offspring; to establish dominance relationships and mating bonds; to persuade reluctant helpers to cooperate; and to discourage thieves, cheats, parasites and even predators. Punishers are generally dominant individuals that can exact retribution at little cost - though subordinates may sometimes wait for favourable opportunities to punish dominant individuals at times when they cannot retaliate effectively.

References

ANDERSSON, M.: *Sexual Selection*, Princeton (Princeton University Press) 1995.

ARNQVIST, G.: "Multiple mating in a water strider: mutual benefits or inter-sexual conflict?", *Animal Behaviour,* 38 (1989), pp. 749-756.

AURELI, F./COZZOLINO, R./CORDISCHI, C./SCUCCHI, S.: "Kin-oriented redirection among Japanese macaques - an expression of a revenge system", *Animal Behaviour*, 44 (1992), pp. 283-291.

BROWN, J. L.: *Helping and communal breeding in birds*, Princeton (Princeton University Press) 1987.

CHAPMAN, T./HUTCHINGS, J./PARTRIDGE, L.: "No reduction in the cost of mating for *Drosophila melanogaster* females mating with spermless males", *Proceedings of the Royal Society of London, Series B,* 153 (1993), pp. 211-217.

CHENEY, D. L. (1983a): *Proximate and ultimate factors related to the distribution of male migration*, ed. by R. A. Hinde, Oxford (Blackwell Scientific Publications) 1983.

CHENEY, D. L. (1983b): *Extra-familial alliances among vervet monkeys*, ed. by R. A. Hinde, Oxford (Blackwell Scientific Publications) 1983.

CHENEY, D. L./SEYFARTH, R. M.: "The recognition of social alliances by vervet monkeys", *Animal Behaviour,* 34 (1986), pp. 1722-1731.

CHENEY, D. L./SEYFARTH, R. M.: *How Monkeys See the World*, Chicago (University of Chicago Press) 1990.

CLUTTON-BROCK, T. H.: *The evolution of parental care*, Princeton (Princeton University Press) 1991.

CLUTTON-BROCK, T. H./DEUTSCH, J. C./NEFDT, R. J. C.: "The evolution of ungulate leks", *Animal Behaviour,* 46 (1993), pp. 1121-1138.

CLUTTON-BROCK, T. H./GUINNESS, F. E./ALBON, S. D.: *Red deer: behaviour and ecology of two sexes*, Edinburgh (Edinburgh University Press) 1982.

CLUTTON-BROCK, T. H./PARKER, G. A. (1995a): "Punishment in animal societies", *Nature,* (1995), pp. 209-216.

CLUTTON-BROCK, T. H./PARKER, G. A. (1995b): "Sexual coercion in animal societies", *Animal Behaviour,* 49 (1995), pp. 1345-1365.

COOK, S.E./VERNON, J.G./BATESON, M./GUILFORD, T.: "Mate choice in the polymorphic African swallowtail butterfly, *Papilio dardanus*, male-like females may avoid sexual harassment", *Animal Behaviour*, 47 (1994), pp. 389-397.

CORDS, M.: *13th Congress International Primatological Society*, ed. by A. Ehara et al., Amsterdam (Elsevier Science Publications) 1991.

DATTA, S.: *Primate Social Relationships*, ed. by R. A. Hinde, Oxford (Blackwell Scientific Publications) 1983.

COERCION

DE WAAL, F.: *Chimpanzee Politics*, London (Allen & Unwin) 1982.

DE WAAL, F. B. M.: *Peacemaking Among Primates*, Cambridge, Mass. (Harvard University Press) 1989.

de WAAL, F. B. M.: *Primate Social Conflict*, ed. by W. A. Mason and S. P. Mendoza, Albany (State University of New York Press) 1993.

EMLEN, S. T./WREGE, P. H.: "Forced copulations and intra-specific parasitism: two costs of living in the white fronted bee-eater", *Ethology,* 71 (1986), pp. 2-29.

ENOMOTO, T.: "Male aggression and the sexual behavior of Japanese monkeys", *Primates,* 22 (1981), pp. 15-23.

FOWLER, K./PARTRIDGE, L.: "A cost of mating in female fruit flies", *Nature,* 338 (1989), pp. 760-761.

GOODALL, J.: *The Chimpanzees of Gombe*, Cambridge, Massachusetts (Belknap Press) 1986.

HAMMERSTEIN, P.: "The role of asymmetries in animal contests", *Animal Behaviour*, 29 (1981), pp. 193-205.

HAMMERSTEIN, P./PARKER, G. A.: "The asymmetric war of attrition", *Journal of Theoretical Biology,* 96 (1982), pp. 647-682.

HORSFALL, J.A.: "Brood reduction and brood division in costs", *Animal Behaviour*, 32 (1984), pp. 216-225.

LEONARD, M.L./HORN, A.G./EDEN, S.F.: "Parent-offspring aggression in moorhens", *Behavioral Ecology and Sociobiology*, 23 (1988), pp. 265-270.

LINDBURG, D. G.: *Mating behavior and estrus in the Indian rhesus monkey*, ed. by P. R. Seth, New Delhi (Today and Tomorrow) 1983.

MANSON, J. H.: "Male aggression: a cost of female mate choice in Cayo Santiago rhesus macaques", *Animal Behaviour,* 48 (1994), pp. 473-475.

MAYNARD SMITH, J.: "The theory of games and the evolution of animal conflicts", *Journal of Theoretical Biology,* 47 (1974), pp. 209-221.

MAYNARD SMITH, J.: *Evolution and the Theory of Games*, Cambridge (Cambridge University Press) 1982.

MAYNARD SMITH, J./PARKER, G. A.: "The logic of asymmetric contests", *Animal Behaviour,* 24 (1976), pp. 159-175.

MCKINNEY, F./DERRICKSON, S. R./MINEAU, P.: "Forced copulation in waterfowl", *Behaviour,* 86 (1983), pp. 250-294.

MENDOZA, S. P.: *Primate Social Conflict*, ed. by W.P. Mason and S. P. Mendoza, Albany (State University of New York Press) 1993.

MESNICK, S.: *Evolutionary Biology and Feminism*, ed. by P.A. Gowaty (Chapman & Hall) in press.

MITANI, J. C.: "Mating behaviour of male orangutans in the Kutai Reserve", *Animal Behaviour,* 33 (1985), pp. 392-402.

MULDER, R.A./LANGMORE, N.E.: "Dominant males punish helpers for temporary defection in superb fair-wrens", *Animal Behaviour*, 45 (1993), pp. 830-833.

NADLER, R. D.: *Reproductive behavior and endocrinology of orangutans*, ed. by L. E. M. d. Boer, The Hague (W. Junk) 1982.

NADLER, R. D.: *Sexual aggression in the great apes*, ed. by R. A. Prentky and V. L. Quinsey, New York (New York Academy of Science) 1988.

NADLER, R. D./MILLER, L.C.: "Influence of male aggression on mating of gorillas in the laboratory", *Folia primatol,* 38 (1982), pp. 233-239.

PACKER, C./PUSEY, A. E.: "Female aggression and male membership in troops of Japanese macaques and olive baboons", *Folia primatol,* 31 (1979), pp. 212-218.

PARKER, G. A.: *Sexual Selection and Sexual Conflict*, ed. by M. S. B. and N. A. Blum, New York (Academic Press) 1979.

REEVE, H. K.: "Queen activation of lazy workers in colonies of the eusocial naked mole-rat", *Nature,* 358 (1992), pp. 147-149.

REEVE, H.K./GAMBOA, J.: "Colony activity integration in primitively eusocial wasps - the role of the queen", *Journal of Behavioral Ecology and Sociobiology*, 13 (1983), pp. 63-74.

REEVE, H.K./GAMBOA, J.: "Queen regulation of worker foraging in paper wasps - a social feedback-control system", *Journal of Behaviour*, 102 (1987), pp. 147-167.

ROBERTSON, A. M.: "Female dimorphism and mating behaviour in a damselfly *Ischnura ramburi*: females mimicking males", *Animal Behaviour,* 33 (1985), pp. 805-809.

ROWELL, T. E.: *Social Behaviour of Monkeys*, Baltimore (Penguin) 1972.

RUBENSTEIN, D. I.: *Ecology and sociality in horses and zebras*, ed. by D. I. Rubenstein and R. W. Wrangham, Princeton, New Jersey (Princeton University Press) 1986.

SMUTS, B. B.: *Sex and Friendship in Baboons*, New York (Aldine) 1985.

SMUTS, B. B.: *Sexual competition and mate choice*, ed. by B. B. Smuts, D. L. Cheney, et al., Chicago (University of Chicago Press) 1986.

SMUTS, B. B./SMUTS, R. W.: "Male aggression and sexual coercion of females in non-human primates and other mammals: evidence and theoretical implications", *Advances in the Study of Animal Behaviour,* 22 (1993), pp. 1-63.

STEPHENS, D. W./KREBS, J. R.: *Foraging Theory*, Princeton (Princeton University Press) 1986.

STRASSMAN, J.E./MAYER, D.C.: "Gerontocracy in the social wasp, *Polistes exclamans* ", *Animal Behaviour*, 31 (1983), pp. 431-438.

THORNHILL, R.: "Rape in *Panorpa* scorpion flies and a general rape hypothesis", *Animal Behaviour,* 28 (1980), pp. 52-59.

THORNHILL, R./ALCOCK, J.: *The evolution of insect mating systems*, Cambridge, Mass. (Harvard University Press) 1983.

TRIVERS, R. L.: *Parental investment and sexual selection*, ed. by B. Campbell, Chicago (Aldine) 1972.

WESTNEAT, D. F./SHERMAN, P. W./MORTON, M. L.: "The ecology and evolution of extra-pair copulations in birds", *Current Ornithology,* 7 (1990), pp. 331-369.

WRANGHAM, R.W.: "Female choice of least costly mates: a possible factor in the evolution of leks", *Zeitschrift für Tierpsychologie*, 54 (1980), pp. 352-367.

Chapter 3

Does the Second Law of Thermodynamics Refute the Neo-Darwinian Synthesis?

BRUCE H. WEBER and DAVID J. DEPEW

I. Introduction: Thermodynamic Prelude

Ever since Darwin, there has been a troubled but changing relationship between two great achievements of nineteenth century science, thermodynamics and Darwinian evolutionary theory. The second law of thermodynamics, particularly in its Boltzmannian statistical formulation, predicts that as entropy increases to a maximum, so will disorder. Eventually, energy differences in the universe will disappear, so that no more work will be possible. Darwin's theory, by contrast, implies the possibility that biological systems can, though not necessarily, increase in ordered complexity over time through the action of natural selection. The increase in heterogeneity that is a hallmark of life does not seem to follow the arrow of time that points toward the heat death of the universe.

William Thomson (later Lord Kelvin) drew Darwin's attention to this paradox in a most concrete manner. Thomson performed a thermodynamic analysis of the dissipation of energy from the sun and the earth to estimate an age for the earth. His result was at least one order of magnitude less than

Darwin had assumed. Darwin reacted to Thomson's estimate by shifting some of the work of evolution from selection to the inheritance of acquired characteristics. At the same time, Darwin wisely expressed skepticism about Thomson's estimate on the ground there was as yet insufficient knowledge about the dynamics of terrestrial processes (Darwin 1872). As it turned out, Darwin was shown to be correct after the discovery of radioactivity and nuclear processes. Even so, evolutionary theorists still felt the tension between a physics that saw the universe running down and an evolution that they assumed was progressive. It is true that some important theorists, such as Fisher and Wright, argued that physical and biological selection acted together in a "two tendency universe" (Turner 1985; Hodge 1992). But the arguments to sustain this position were not available to them. One purpose of this paper is to show how different the situation now appears to be.

Erwin Schrödinger addressed the thermodynamic/evolution paradox in his seminal book *What is Life?* Schrödinger sought to reconcile physics and biology by describing not only how order could come from order (replication of the molecules bearing genetic information) but how order could come from disorder (evolution of biological structure in the face of the second law) (see Schneider/Kay 1995). The former issue formed the conceptual inspiration of molecular biology (Watson/Crick 1953); the latter resulted in the development of nonequilibrium thermodynamics (Prigogine 1955). With regard to the thermodynamic/evolution paradox, Schrödinger argued that the second law requires only that the universe as a whole must show an increase in entropy, but does not prevent eddies of order, or negentropy, from existing. A cell, an organism, or even an ecosystem can maintain internal structure, and even become more complex over time, as long as the environment of such structured systems is coupled to them in such a manner that the entropy of the environment increases by an amount greater than the decrease of entropy within the system. For this to occur the system has to remain far from equilibrium. Schrödinger stated:

> It is by avoiding the rapid decay into the inert state of "equilibrium" that an organism appears so enigmatic.... Everything that is going on in nature means an increase of the entropy of the part of the world where it is going on. Thus a living organism continually increases its entropy – or, as you may say, produces positive entropy – and thus tends to approach the dangerous state of maximum entropy, which is death. It can only keep aloof from it, i.e. alive, by continually drawing from its environment negative entropy.... What an organism feeds upon is negative entropy. Or to put it less paradoxically, the essential

> thing in metabolism is that the organism succeeds in freeing itself from all the entropy it cannot help producing while alive. (Schrödinger 1944, pp. 71-72)

Prigogine was one of the first to build upon Schrödinger's resolution of the paradox. However, Prigogine went further to assert that the second law is not merely consistent with evolution, as Schrödinger maintained, but actually helps explain it through the relationship between thermodynamics and self-organization. Self-organization arises in the real world only when specific physical and chemical conditions allow nonlinear kinetic pathways to operate. These conditions require, most importantly, that the system be stabilized far from thermodynamic equilibrium. While the universe, as any isolated thermodynamic system, must increase in entropy, open systems, such as biological cells and organisms, can decrease internal entropy by pumping matter and energy into themselves. This influx can be used to do physical and chemical work, to build structure and to cycle matter and energy internally, while dissipating degraded energy (entropy) to the surroundings of the open system. Such internal structures are often the preferred means for dissipating degraded energy and serve also as internal sinks. They are called "dissipative structures" by Prigogine. Open systems that exhibit dissipative structures include such disparate things as tornadoes, cells, organisms, ecological communities, and possibly economic systems. Dissipative structures collapse energy-matter gradients (Schneider/Kay 1995). If these gradients are not replenished, the structures themselves collapse. But if gradients are sufficiently stable, dissipative structures are able to continue to pull into themselves the resources to maintain themselves in a steady state that remains far from equilibrium. The inherent tendency of open systems to increase, perhaps even to maximize, their dissipative rate is linked to their ability to build better dissipative pathways in the form of more efficient internal structures (Wicken 1987; Swenson 1989, 1997).

The most effective way of building structure and dissipating entropy is by means of autocatalysis. A chemical reaction that produces a substance that can help the production of more of itself will show a rapid amplification of the concentration of that substance. This is called an autocatalytic cycle. Such autocatalytic cycles show nonlinear dynamics (Ulanowicz 1989). By the very nature of the self-organizing processes that produce them, dissipative structures are highly sensitive to initial conditions. Depending on what initial and boundary conditions obtain, they are capable of generating dynamics that, displayed in phase space, produce either order, chaos, or complex organization. Moreover, to the extent that dissipative structures can be influenced by selection, there will be selection pressure for increased stability, thermodynamic effectiveness as measured by entropy production, and catalytic efficiency in comparison to other dissipative structures that are competing for resources.

II. What is the Role of Natural Selection in the Face of Thermodynamic Constraints?

The question posed by the title of this paper is whether the nonequilibrium thermodynamic perspective on living things represents a refutation of Darwinian and neo-Darwinian concepts of natural selection as sources of biological order. We have answered this question in the negative in previous writings (Depew/Weber 1988, 1995; Weber/Depew 1996). We have argued that, from its inception, the Darwinian research tradition, has appealed to systems dynamical models to interpret its basic concept of natural selection. When it faced a major challenge from the rise of genetics, for example, it was able to achieve a new evolutionary synthesis by changing from a Newtonian dynamical background picture to a Boltzmannian one. The issue today is whether an appeal to the emerging field of complex systems dynamics, in which self-organization occurs under thermodynamic constraints, will similarly invigorate the Darwinian tradition, or alternatively, lead to its demise. The type of thermodynamically-informed Darwinism that we espouse, which has appropriately been called "ecological Darwinism," does not reject the insights of neo-Darwinism. Although it proposes a revision of its conceptual framework, it allows us to view natural selection in a richer light and to have a better understanding of the conditions of its effectiveness.

We will explore some of the implications of this view by addressing three interconnected questions. First, if biological systems at various levels of integration have self-organizing properties by virtue of thermodynamic constraints and nonlinearities, how can natural selection retain its central role in accord with our assertion that natural selection is the conceptual core of the Darwinian research tradition (Depew/Weber 1995)? Second, can natural selection and self-organization work in conjunction with each other? If so, under what conditions does this occur, and, what sort of entities does selection select? Finally, if we succeed in reconceiving natural selection and self-organization as mutually supporting and mutually entailing processes (Depew/Weber 1995; Weber/Depew, 1996), can this argument be extended to behavioral cycles, and hence to cultural and economic phenomena?

Before proceeding, it should be noted that there exists an alternative evolutionary research tradition to Darwinism that we dub "evolutionary developmentalism," which existed before Darwin and still thrives today (Ho/Saunders 1984; Goodwin/Saunders 1989; Salthe 1993). The main theme of the evolutionary developmentalist tradition is that autonomous ontogenetic, ecological, and phylogenetic dynamics, *rather than selection*, are the creative forces in the hierarchical complexification of life. The recent emergence of thermodynamically-driven self-organization has emboldened adherents of this tradition (see Brooks/Wiley 1986, 1988; Swenson 1989, 1997; Ho 1993; and Salthe 1993 as examples of different ways of grounding this tradition in physical law). In its contemporary incarnation, the "process structuralism" of Brian Goodwin, for example, gives the self-ordering properties of morphogenetic fields a causal power that directly rivals the power of natural selection (Goodwin 1994). Mae-Wan Ho goes further and in fact sees little substantive role in evolution for natural selection as opposed to developmental processes (Ho 1984, 1988). In this view, species are instantiations of universal "laws of form" and so are natural kinds (Thompson 1917, 1942; Goodwin 1994) rather than the historically unique particulars that are produced by the action of natural selection, which are explainable only through narrative reconstructions of unique sequences of events (Gould 1989; Lewontin 1991). The *possibility* of natural selection is not denied by most evolutionary developmentalists. But it is relegated at most to a supplementary role of weeding out the unfit and modulating evolutionary rates (Brooks/Wiley 1986, 1988). Developmental evolutionists reject any version of Darwinism, then, however expanded and thermodynamically or ecologically informed, that does not hold to the integral vision of self-moving evolutionary complexification at various levels of scale (Salthe 1993; Swenson 1989, 1997; Swenson/Turvey 1991). Developmental evolutionists tend, in addition, to believe that no Darwinian theory can possibly meet these conditions. Stanley Salthe, for example, combines hierarchy theory, systems ecology, and nonequilibrium thermodynamics to affirm strong isomorphisms between ontogeny, ecological succession and phylogeny (Salthe 1985, 1993). The only sort of systems that can evolve are those that develop. Anything that develops, however, might also evolve. Thus Salthe concludes that evolution itself is a developmental process. For his part, Rod Swenson sees this process as instantiating a universal law of development-by-differentiation of the sort that was anticipated, but not explained, by Spencer's expansion of Von Baer's law of ontogeny (Swenson 1989, 1991a,1991b, 1997).

The contemporary developmental evolutionists cited above have made good use of the resources of nonequilibrium thermodynamics, nonlinear dynamics, and complex systems dynamics generally. Our argument is that Darwinians can also utilize such resources, and may even revitalize their conceptual core of natural selection by using such dynamical models (Depew/ Weber 1995).

Let us begin with a closer consideration of what living systems are. Thermodynamically, living systems belong, as we have seen, to the general class of dissipative systems. But they differ from purely physical and chemical dissipative structures (such as Bénard convection cells and the chemical wave patterns of the Beloussov-Zhabotinsky reaction) in that cells, for example, do not have to reinvent metabolic pathways in each cycle or generation, since they carry within themselves the information that encodes the organization of metabolism. Genetic information records the structures that have proven useful in the past for producing the dissipative structures needed for survival and for passing that crucial information down to the next generation. For this reason, we have defined living systems thermodynamically, as bounded, informed, self-replicating, dissipative autocatalytic cycles (Wicken 1987; Morowitz 1992, Depew/Weber 1995).

Such a thermodynamic definition gives insight into the origin of life (Depew/Weber 1995; Weber, in press). Protocellular entities at the dawn of life, when nucleic acids and proteins were coupled synergistically into autocatalytic cycles, can be expected to have progressively acquired more precisely defined metabolic functions over time (Wicken 1987; Weber et al. 1989; Weber 1991; Weber, in press). Energy-capture and phosphorylation mechanisms acting across proto-membranes would have driven polymerization of amino acids and nucleotides in a chemically bounded space to produce "generic" proteins and nucleic acids. These macromolecules can be expected to show weak but broad catalytic activity. Poised away from equilibrium, nonlinear interactions and autocatalytic cycles would spontaneously arise through the action of these catalytic polymers. A proto-metabolism would have emerged through a process of catalytic closure (Kauffman 1993, 1995). In this early stage of proto-cellular evolution, there would not be much information in the usual genetical sense. However, through computer simulations of the process of catalytic closure Kauffman has proposed that there is a type of mutual, metabolic information that can be passed on to "daughter" protocells. The *ensemble* of protocells would have exhibited weakly heritable variation. With catalytic closure, moreover, there would have arisen increased rewards for catalytic efficiency, and thus selection pressure for a better fit between the sequence space of a given macromolecule and the catalytic task space with which it roughly overlaps. Even though there would be no genetic information yet, there would have been competition with other such autocatalytic systems for energy fluxes. As successful patterns of dissipation emerged, and as protocells evolved into true cells, a progressive tightening of the nucleic acid-protein relationship would be expected to have taken place.

This competition was, at first, neither the type in which entities are sorted by physical selection for stability properties, nor was it yet biological selection for reproductive fitness. It was thermodynamic or chemical selection for effective entropy production and catalytic efficiency, in which the relevant units of selection are energy-capturing and energy-utilizing autocatalytic cycles. While self-organization of autocatalytic dissipative systems would have provided a source of order and ordered units, therefore, selection for thermodynamic effectiveness would have strongly shaped that order, and the sorts of beings on which it could operate. Indeed, the boundary between life and nonlife would not have been crossed unless and until selection became the predominant process. At the same time, selection pressure so construed would have favored any entity that increased the efficiencies of its autocatalytic cycles by storing and stabilizing the information needed for autocatalysis. There would have been an enormous competitive advantage for those entities that could "remember" information that enhanced autocatalytic activity by encoding it in the polymers of nucleic acids formed by chemical selection itself. In addition to internalizing metabolic information, such entities would have had to be able to pass this information on to successor entities, for the ability to do this in one generation alone would not count as storing and stabilizing information. Indeed, at this stage, the ability to reproduce and pass on the relevant information would have become more important than just storing and deploying metabolic information. In our view, fitness is a measure of this ability, and natural selection of the reproductively fit is an emergent process from chemical selection for the autocatalytically efficient and, more distantly, from the physical selection for the stable.

This is to say that natural selection cannot be reduced to chemical or thermodynamic selection. It is certainly not to say, as Richard Dawkins does, that natural selection is just a case of physical selection: Dawkins is wrong to assert that the "earliest form of natural selection was simply a selection of stable forms and a rejection of unstable ones" (Dawkins 1976, p. 13). Natural selection is for the reproductively fit whereas physical selection is for the stable, and chemical selection for the catalytically efficient. This does not imply that physical, chemical and natural selection are not parts of a single coherent process. On the contrary, it is to say that evolutionary processes exhibits emergent levels, emergent properties, and emergent phenomena. The problem of the origin of life is best construed, accordingly, as explaining rather than presupposing the emergence of the phenomenon of natural selection (Depew/Weber 1995; Weber/Depew 1996; Weber, in press).

III. Development and the Units of Natural Selection

In *Darwinism Evolving*, we argued that there are two strands of genetical (or neo-) Darwinism: an 'English wing,' growing out of the work of Ronald Fisher, which is committed to context-independent fitness values of individual genes, and an 'American wing,' inspired by Theodosius Dobzhansky, which stresses the principle of genetic relativity in constantly changing environments (Fisher 1930; Dobzhansky 1937; Depew/Weber 1995). In the latter view, there is a selection pressure for mechanisms that maintain variation. For this reason, Dobzhansky saw heterosis as inherently adaptive. He hypothesized, moreover, that adaptability – a higher-level trait that creates and maintains variation – is itself an evolutionary adaptation. Accordingly, whereas the Fisherian wing has become gene centered and reductionist, according causal efficacy to "selfish genes" (Williams 1966, 1992; Dawkins 1976; Dennett 1995), some members of the Dobzhanskyian wing of the Modern Evolutionary Synthesis have argued for an "expanded synthesis" with selection operating at a number of levels, complemented by the working of chance and self-organization, and non-adaptationist explanations (Lewontin 1974; Gould/Lewontin 1979; Gould 1982; Eldredge 1985, 1995). In this context, it was not unexpected when John Maynard Smith, the elder statesman of British Darwinism, reacted strongly against applying self-organizational and thermodynamic models to evolutionary problems (Maynard Smith 1995). For those familiar with or working within the American Darwinian tradition, however, the choice between natural selection and self-organization does not appear nearly as stark. Self-organization may well be a threat to the main line of British Darwinism. But it is likely to extend rather than undermine the American wing of the Modern Synthesis. Thus it is no surprise that complex systems dynamics has been applied within the more pragmatic American Darwinian tradition to questions of ecosystem behavior (Wilson 1980; 1992, in press), to the problem of the relationship between natural selection and self-organization (Kauffman 1993, 1995), and to questions about co-evolution within ecosystems that may themselves exhibit self-organization as well as dissipation (Eldredge 1985, 1995).

In contrast to the view of organisms as passive collections of disaggregated traits and of genes coding for them, and especially to Dawkins' vision of organisms as ephemeral gene-reproducing vehicles, the American wing of the Modern Evolutionary Synthesis has stressed the notion of organisms as active creators of their world. As Richard Lewontin writes, "The fitness at a single locus ripped from its interactive context is about as relevant to real problems of evolutionary genetics as the study of the psychology of individuals isolated from their social context is to an understanding of man's sociopolitical evolution" (Lewontin 1974, p. 318). DNA, after all, does not replicate itself. Nor can it make protein by itself. These activities are possible only within a cellular, metabolic, and developmental context, that is, in a context of dissipative energetics and information retention. Within this actual biological context, it is not merely false but incoherent to state that the genes contain all the information that instruct otherwise passive and inert cellular machinery. Since the molecular revolution in biology, a cybernetic model cast in the language of senders, receivers, codes, decoders, and programs has led us astray about this issue. Development becomes the expression of the information in the cybernetic program of the genome. This conception is not even metaphorically plausible, however, unless "the encoded representation of the phenotype (or of the instructions for building it) is thought to reside in the genes," as Susan Oyama (Oyama 1985, p. 26) put it – a notion that derives whatever meaning it has from remnants of the preformationism into which "gene-talk" has always been prone to fall. Instead, the best way to define the information stabilized by genes is to characterize it in terms of biological function (Oyama 1985). The implication for developmental biology is support for the traditional rival of preformationism, epigenesis, in which faithful reproduction is part of the process of progressively articulated development, and thus is causally a genuinely new creation (without being *creatio ex nihilo,* of course) in each generation and at each phase in the developmental process itself. From this perspective, it is at least as informative to speak of genes as catalysts for a complex set of cellular interactions, stabilizing them and facilitating them as catalysts do, as to speak of genes as causally primary (Weber *et al* . 1989). To use the cybernetic information metaphor correctly, we can view genes as channels through which the organism transmits information about developmental factors–a view that is much more realistic than the rival notion that development is the channel for transmission of the genetic information.

On this view, each species has a species-specific set of developmental resources that are expressed in a self-organizing process, which does not necessarily need a central information source to produce species-specific traits (Oyama 1985). These resources include genetic information. But they also include cytoplasmic and metabolic structures, as well as information in the zygote, behavior, environmental factors, and even social context. For Oyama, then, "Ontogeny is the contingent functioning of entire developmental systems. It is the systems, mobile networks of organism-niche relations, that are 'transmitted' (reconstructed) in heredity" (Oyama 1988, p. 258). Information is not just transmitted from the genes to the phenotype. Rather the resources needed to complete the latter stages of development are produced in the earlier stages, so that functional information is a result of ontogeny.

From this perspective we can agree with one point made by evolutionary developmentalists: Only developmental entities can biologically evolve. Yet we do not think that it follows from this that evolution is itself a developmental, rather than a selectionist, process. Our thought, which differs from that of most Darwinians, is that natural selection can select only among entities and processes that come to be through a developmental process. That is, we think that natural selection is not entirely neutral about the sorts of things it can range over and, therefore, that Lewontin's famous minimal definition of selection as any process for which there is differential retention of heritable variation is too broad. More particularly, we assert that natural selection is to be seen as acting not on genetic programs, or even on organisms as usually understood, but rather on variations among developmental processes or *life cycles.*

This view of selection has recently been powerfully articulated by a number of workers, such as Paul Griffiths and Russell Gray, who argue that there is a rich array of developmental resources associated with life cycles, that these form a complex system that is itself replicated in development, and that differential replication of variants among life cycles is the most perspicuous way to construe paradigmatic units of selection (Griffiths/Gray 1994). "The developmental system goes far beyond the traditional phenotype, yet all its elements are parts of the evolutionary process" (Griffiths/Gray 1994, p. 278). The developmental system is not, on this view, a collection of separately evolved and aggeggated features, but rather an integrated, non-decomposible, but quite variable whole that is itself the product of many evolutionary processes over phylogenetic time. For Griffiths and Gray "an evolving lineage... [is] a series of cycles of a developmental process" (Griffiths/Gray 1994, p. 291), in which the developmental process, the central theoretical entity, is made up of developmental interactions with developmental resources that have played a role in the evolution of the process. In each variant life cycle, the functional characteristics of the lineage must be reconstructed anew from the resources available. "The evolution of complex, functional structures requires a repeated life cycle during which structures are repeatedly reassembled" (Griffiths/Gray 1994, p. 293). Such a perspective eliminates the need for the replicator/interactor distinction because living systems, which are interactors, replicate themselves only in the presence of the full array of interactive developmental resources. The replicator/interactor distinction is an artifact of too neutral, indeed too mechanistic a conception of natural selection.

Even though it requires a significant reformulation of evolutionary theory, this approach expands the explanatory power of evolutionary theory. In its emphasis on the embeddedness of entities in energy-processing cycles upon which selection can act, moreover, it agrees well with the causal explanations of the thermodynamical formulations of evolutionary theory, which rough-hew the sorts of partially self-organizing entities that live, vary, develop, and biologically evolve. Still, the developmental systems approach, as articulated by Griffiths and Gray, retains the centrality of the concept of natural selection as a process of sorting variation to produce differential reproduction, even as it changes our notion of the entities upon which natural selection acts. Variant life-cycles compete and have fitness values. But fitness is measured not against an independent environment, nor solely as gene-frequency changes, but as the comparative self-replicating power of the whole system. "Evolution," says Oyama, "is change in the distribution and constitution of developmental (organism-environment) systems" (Oyama 1988, p. 256). While this does not deny that evolutionary change can be tracked in part by alteration in gene frequencies, it does open to consideration a richer conception of causally important roles for functional information from across the life cycles of organisms (Oyama 1993).

IV. Development and Evolution in Ecosystems

In contrast to the developmental systems theorists discussed above, some evolutionary developmentalists consider the strong isomorphies between the pattern of development in organisms and succession in ecosystems to be evidence that ecosystems are living entities and that they evolve. We do not agree that ecosystems, *qua* ecosystems, are alive. Ecosystems are composed of highly integrated living things, but they are themselves not living things. We do agree that ecosystem phenomenology exhibits dynamical patterns of succession that resemble those of development (Ulanowicz 1986). Since we do not believe, however, that biological evolution is, as such, a developmental process, we do not think that ecosystems *evolve*. Succession may well be due to a type of chemical or thermodynamical selection acting upon autocatalytic cycles of energy-matter flows within ecosystems, producing self-organization under thermodynamic constraints and imperatives (Schneider 1988; Weber et al 1989; Depew/Weber 1995). Admittedly, organisms and groups of organisms act as nodes through which energy-matter transfers occur within ecosystems (Ulanowicz 1986; Wicken 1987). But only *lineages* of those entities that are components of ecosystems can biologically evolve, and their evolution is to a considerable extent a function of natural selection. Still, the very fact that the ecological systems in which organisms and groups are embedded are hierarchically self-organizing suggests why natural selection can be expected to be a multi-level process, as the expanded Darwinism of Dobzhansky's heirs has suggested..

Population ecologists have recognized the complexity of interactions of species within communities, which often produce dynamics best described by the use of chaotic attractors (Segal/Jackson 1972; May 1974, 1978, 1987; Ornstein 1989; Drake 1990; Strauss 1991). The variation produced by sensitivity to initial conditions in such complex systems can provide a source of variation upon which selection can act (Wilson 1992). In general, however, neo-Darwinians have been reluctant to the countenance a role for self-organization in evolution (see for example Maynard Smith 1995, Wilson 1996).

Their argument is that self-organization can give rise to pattern, but not to the functional organization that, in their view, only natural selection achieves. We have argued above that we would not expect self-organization to work alone to produce biological order or functional organization. Even at pre-biotic levels, some sort of selection is required. We do claim, however, that, since evolution is co-evolution within communities, self-organization and natural selection working together can be more effective than natural selection alone in producing functional organization. Complex systems dynamics allows us to discern the conditions under which selection can be most effective in producing adaptations. The question is whether the accounts of functional organization given by thermodynamic/ecological Darwinism and those given by orthodox neo-Darwinism are compatible.

The notion that functional organization at the level of communities is due to selection discriminating among entities other than organisms or genes has been espoused by some neo-Darwinians (for example, Wilson 1980, 1992, in press; Wilson/Knollenberg 1987; Wilson/Sober 1989, 1994), even though the idea of group selection at levels above individuals had been severely criticized, along with the discredited notion of "superorganisms" (Williams 1966). In this context, for example, George Williams proposed that the gene is the fundamental unit of selection, and that individual organisms and groups are merely the environments of genes. Selection is strongest, on this view, at the level of genes. Above the level of individuals, it may be conceptually and empirically possible, but too weak to produce adaptations. Yet David Sloane Wilson and Elliott Sober (Wilson/Sober 1989, 1994), agreeing with Hull and Brandon (Hull 1988; Brandon 1990) draw upon Dawkins' distinction between replicators and vehicles of selection to defend higher units of selection. What is important about vehicles is that the fate of gene replicators is joined to the fate of the vehicle. Genes within an organism may have different fitness values, but their fate as a whole is dependent upon the fate of the individual in which they are contained. Generalized, this concept can be applied to any level of the biological hierarchy. If organisms within a group do not differ in fitness simply because they share the fate of all other members of the group, then selection is not acting upon them, but, if at all, upon their vehicle. They are all in the same boat, to adapt Dawkins' rower metaphor (Dawkins 1976). There is, then, a trichotomy of levels: that which is contained within the vehicle (analogous to an organ in an individual organism), the vehicle itself (analogous to the organism as traditional object of selection), and the population of vehicles. Genes are rarely vehicles (one exception is the gene of the parasitic wasp *Nasonia vitripennis*, which destroys all the other genes of the male wasp to ensure its transmission (Werren 1991). Most of the time the vehicle will be the individual organism. It is not only theoretically possible, however, but empirically the case, that there can be vehicles that are demes or subsets of demes. Insect colonies and phoretic insects, such as mites that live on and are transported by beetles, are examples. An increase of fitness of mites on a given beetle that decreases the fitness of the host beetle will decrease the fitness of the phoretic associates. If we look at the population of mites on a single beetle from the conventional perspective, we would expect selection to favor mite fitness, in spite of the cost to the beetle. Conversely, decrease of fitness of the mites through mutualistic behavior that increases the fitness of the beetle would presumably be selected against. (This is analogous to the problem in sociobiology of the selection against 'altruistic' behavior.) But in reality there are many beetles, each with its isolated population of mites. Each mites-beetle association represents a "community," and if some communities promote the survival of the beetle better than others, the differential survival of communities will select for mutualistic behaviors that would otherwise be selected against within a given community. Thus there is a balance of selective forces between and within communities that is dependent upon population structure (as a long succession of American neo-Darwinians going back to Sewall Wright has held). These empirical results are consistent with a theoretical model that countenances one form of group selection (Wilson/Knollenberg 1987). This model is easily generalizable to communities and ecosystems (Wilson 1992, in press).

The idea that functional organization and adaptation within ecosystems is due to natural selection acting on "trait groups" is consistent with our version of thermodynamical Darwinism, since we do not conflate the self-organizational developmental patterns exhibited by ecosystems with functional organization of the living entities that are embedded within the system. Natural selection sorts organisms and trait groups, *within a thermodynamically described and structured ecosystem,* to produce the evolution of lineages. So true is that that evolution by adaptive natural selection can *destroy* the developmental pattern observed in ecosystems. When organisms are selected that draw more energy-matter through themselves, they increase the flow of energy-matter through the system as a whole. But (as Depew/Weber 1995 and Weber/Depew 1996 failed to acknowledge) increased flow can disrupt the system unless it can be accommodated by the other nodes within the system, only in which case the system as a whole appears to obey Lotka's laws. The organism, therefore, does not evolve *to* enhance the energy-matter flow of the system in which it is embedded, although such a contingent consequence can indeed occur under certain conditions. The embeddedness of the objects of adaptive natural selection in thermodynamically-dissipative ecosystems is, then, a necessary but not sufficient condition for biological evolution. It explains why the entities that are selected are partially self-organizing, dissipative structures, why they fall into a series of hierarchical levels, and why evolution takes place in a co-evolutionary theatre. But it does not by itself account for their evolution or their functional roles.

V. Can Natural Selection and Self-Organization Work Separately?

We have postulated a complex interplay between self-organization and natural selection in developmental processes and ecosystem dynamics. We will now examine this relationship at a more abstract level. Stuart Kauffman's NK model permits a very general description of any system that consists of N components with K interactions among the components, and in which there can be any number of states for each component. N can represent or model the number of amino acids in a protein, the number of polypeptides in a catalytic ensemble, the component genes of an ontogenetic program, the number of genes in a population, and a good deal else. These models are mapped onto a fitness landscape, where fitness does not need to be defined absolutely, but only as relative fitness. Variation of parameters shows three classes of dynamical behavior. When N is large and K is also large, the fitness landscape is rugged, with peaks about equal to the mean fitness of the ensemble. The dynamics show sensitivity to initial conditions and large, strange attractors, both of whch are characteristics of chaotic systems. When K=0, there is a smooth fitness landscape, with only one peak of fitness that is much higher than the ensemble mean fitness. This condition corresponds to that assumed by Fisher, in which a change in one component makes only a small change in fitness and does not affect other component. Around K=2, a number of high, adaptive peaks occur near each other in a region Kauffman calls the "edge of chaos" and Robert Ulanowicz perspicuously calls the "window of vitality" (Ulanowicz 1995). This corresponds to the type of adaptive landscape employed by Sewall Wright (Wright 1932). Perturbations in this region usually have small effects. But occasionally there can be a major rearrangement of relations between the components. The shape of the fitness peaks in this region reflect ensemble properties and the nature of the interactions between the components. It is in this region alone that adaptive behavior is possible. This situation seems to characterize complex adaptive systems of any sort (Gell-Mann 1995). At this high, abstract level, Kauffman sees selection as a co-creating partner with self-organization. He proposes a new and deeper role for natural selection from this perspective than Darwinians have thus far envisioned. It is selection itself that pulls systems into the region of adaptive behavior that emerges at or near the interface of the ordered and chaotic regimes:

> Ordered systems encounter relatively correlated landscapes, chaotic systems adapt on very rugged landscapes. Landscape structure governs both evolvability and sustained fitness...Parallel processing systems lying in this interface region between order and chaos may be the best able to adapt and evolve. Further, *natural selection* itself may be the force which pulls complex adaptive systems into this boundary region... It seems plausible that the most complex, integrated, and also the most evolvable behavior might occur in the boundary region near the edge of chaos (Kauffman 1993, pp. 218-219, original emphasis).
>
> If it proves true that selection tunes genomic systems to the edge of chaos, then evolution is persistently exploring networks constrained to this fascinating ensemble of dynamical systems. The generic properties of this perhaps most important ensemble emerge as the best hypothesis to account for the remarkable order in organisms. If this overall view proves useful, then the manifold marriage of self-organization and selection consists in constraint of genomic systems to the well-wrought ensemble as a result of selection's achieving systems best able to adapt... (Kauffman 1993, p. 522).

In this model the action of natural selection is to favor those entities that the interaction between self-organization and selection has produced, which are generally more fit, and beyond that to maintain such systems in the region of phase space that affords them the greatest opportunity of further evolution by generating and retaining variation. Thus, as Dobzhansky had argued a half a century earlier, adaptability is itself an adaptation.

When Kauffman brings his general approach down to real cases, he either explicitly or implicitly assumes that the systems are far from thermodynamic equilibrium and have nonlinearities. Thus his conclusions are compatible with those that, like our own, emphasize concrete thermodynamic imperatives and constraints as causal conditions for the evolution of complex adaptive systems. In this view, natural selection, which emerged with life itself, has an evolutionary history. Rather than simply being a model of a natural process, it is an evolutionary phenomenon in its own right (Weber/Depew 1996). At the very least, this means that stochastic, selective, and self-organizational processes are empirically, causally, and perhaps even conceptually intertwined in the evolution of the living subclass of developing systems. Natural selection has a very important, indeed an indispensable, role to play, both in favoring evolvability, and in selecting between the variation that chance and self-organization make possible. But it does not have to do all the work of producing biological order. At the same time, although self-organization plays a very important role in this theory, without the action of selection systems would likely stay in the chaotic regime, and would have a bleak evolutionary future.

We do not deny that natural selection acts upon heritable variation, even if not all the inheritance is necessarily in the genes. But if the replicator/vehicle distinction were to be relaxed, in accord with thermodynamic assumptions, greater consistency might arise between Wilson's hierarchical selection of vehicles and the developmental-systems approach. Vehicles and groups would be viewed as components in the range of resource in a nested hierarchy. Mutualism and cooperation can have thermodynamic, selective, and developmental rewards in the face of competition. What is more, a fully naturalistic sociobiology may be possible that avoids the well-known problems of gene-centered, reductionistic accounts of adaptation.

VI. Implications for Sociobiology

We have argued above that the thermodynamic and ecological perspective fits congenially with that of the developmental systems theorists. Both can be made consistent with the expanded notions of selection acting at various hierarchical levels. All of these approaches, moreover, fall within the Darwinian research tradition as we have defined it (Depew/Weber 1995). None refutes the Modern Evolutionary Synthesis, or neo-Darwinism. However, one particular strand of the Modern Evolutionary Synthesis, which derives from

Fisher and has provided the conceptual foundation for much of the sociobiology espoused by such figures as Dawkins and Daniel Dennett, is quite hostile to these redescriptions of the action of natural selection. Dawkins' replicator-centered reductionism seems to us to be a thin platform upon which to construct a sociobiology. We would rather build a sociobiology upon the idea that evolution acts upon ecologically-embedded yet differentially varying life cycles encompassing developmental-informational resources. What would be the consequences of doing so?

In English-speaking countries, the term 'sociobiology' refers both to a field of study, and to a specific theory within that field. The field is the biological bases of social behavior in animals as well as in humans. Konrad Lorenz was a sociobiologist in the first, disciplinary sense, but not in the second. Indeed, much anglophone sociobiology has been explicitly devoted to refuting Lorenz' approach in favor of a genocentric version of Darwinism. Thus in a cultural milieu in which cooperation, and *a fortiori* self-sacrificing altruism, have long represented an anomaly for the dominant Darwinian styles of thinking, Darwinian sociobiologists have proposed to explain social behavior not by transferring the unit and beneficiary of natural selection upward to the group, and certainly not to the species, but downward to the genes. To the extent that phenotypes, vehicles, or interactors exist for the sake of getting genes or replicators better represented in the next generation, they say, there seems no reason why natural selection cannot, under specific conditions, favor cooperative, and even self-sacrificing behaviors. One might as easily die, as Haldane once averred, for eight cousins as for two siblings, since the same average complement of genes will be transmitted to the next generation in either case. On this view, the selfishness that is at the core of the Darwinian tradition is not cancelled, but preserved by transferring selfishness from the organism to the gene.

What does our account imply about this claim? We argue that natural selection arises in and is restricted to entities that undergo a developmental process. The units of natural selection are reliable, repeatable, but varying cycles in which morphological and behaviorial traits develop. There is, in our view, no reason why patterns of behavior embedded in proto-cultural or cultural fields cannot count as units of selection so considered. On the contrary, there is every reason to think that they can become potentially fitness-enhancing, or adaptive, aspects of total phenotypes that distinguish one sub-population from another.

In a colony (deme) of Japanese macaques, for instance, one individual learned efficiently to separate wheat from chaff by throwing wheat on water. This behavior, which has an adaptive advantage, was learned by other members of the colony. More importantly, this adaptive behavior has become part of the unique culture of this colony and has been passed down through a number of generations (Boyd/Richerson 1985; Brandon 1990). While this behavior does not have a genetic basis, it is, in fact, part of the developmental resources of the life cycles of individuals in the colony. Over time this deme may have a selective advantage over other competing demes, in accord with the model of group selecton proposed above. Of course, this assumes learning abilities, in macaques as well as humans – learning abilities that themselves are the result of previous adaptive evolution. There is clearly important information in the human genome, for example, that makes possible human language. But the expression of speech and abstract language skills depend critically upon contingencies of the total developmental process, which include continuous exposure to spoken language during childhood. Accordingly, human behavior, learning, and culture are a part of the developmental processes and life cycles of humans upon which selection, chance, and self-organization act. By undercutting the persistent antinomy between genocentric and and overly morphologically-centered conceptions of nature, on the one hand, and on the other overly Lamarckian conceptions of cultural evolution, this way of looking at the matter makes it incoherent to draw a line between "nature and nurture" (Oyama 1985, 1992, 1995).

A second theme of genocentric sociobiology arises from the fact that social animals evolve communicative and technological devices – what Dawkins called "extended phenotypes" – as well as the behaviors required to use them (Dawkins 1982). Tool-use involves ideation and communication. Thus Darwinian sociobiologists have tried to find units of communication that are propelled through a given society by selection pressures that are conceived as closely analogous to the units of natural selection as possible. Hence arises pervasive talk about "memes."

In our view the unit-character of ideas that constitute "memes," which are propelled through society, arises from too strong analogy between genes and memes. Genotypes, by this very Dawkensian notion, don't develop. Like units of currency, they only increase or decrease their representation in a population. In this spirit, David Hull has described the growth of scientific knowledge as a competition among memes, as culture, as it were, red in tooth and claw (Hull 1988). But ideas do not have such a unit character.

Rather, like the poems that T.S. Eliot discussed (Eliot 1960), ideas are embedded within complex, ever-changing traditions, and are certainly defined and redefined within, or against, those traditions as they develop. New ideas are in fact perturbations of old ideas, becoming parts of a revised tradition that is then passed down to future generations for further emendation. Only a strong reorientation of Darwinism to the develomental matrix of evolutionary dynamics can properly connect natural selection to social change.

Consider the "meme" of natural selection itself, for example. Dennett has claimed that the concept of natural selection is itself a meme with high fitness value (Dennett 1995). By contrast, we have asserted that the idea of natural selection is not a constant idea, whose representation in the environment changes, but a changing idea that develops along with the tradition of which it is a part (Depew/Weber 1985, 1995). New reconceptualizations of natural selection based on thermodynamics, hierarchical levels of selection, and developmental systems, for example, may become incorporated within the life cycle of the culture that sustains the Darwinian tradition, and in turn may provide a more adequate conception of cultural dynamics. We view this conception as metaphorically much closer to a notion of nested life cycles as units of selection than to genic reductionism.

It virtually follows from this that the abuses of a specific theory of human sociobiology are not sufficient to show that sociobiology as a field is, in and of itself, an abuse of biology. It is important to ground sociobiology, however, in our best biology. The promise of the thermodynamically-, developmentally-, or hierarchically-informed alternative descriptions of evolution for sociobiology is that a richer and more realistic biological perspective can capture more of the richness and complexity of actual human culture. To focus only on a selfish-gene's perspective is not only short sighted, but actually ignores much, if not most, of the important phenomena in human sociobiology. Group-level functional organization in human society is no more a product of self-interest than it is for the social insects (Wilson/Sober 1994). Not only does adaptive human behavior reflect the balance of selection at various levels, but human cultural norms of behavior can affect that balance and the functional organization that results. Our moral, and immoral, behavior becomes part of a total life cycle that, in its variants and group-level adaptedness, affects the evolution of our kind.

Our point is that we should not trim down our social theory, with its descriptions of the rich and complex patterns of meaning of human life, to fit one particular strand within the neo-Darwinian program, but should instead seek an evolutionary theory within the Darwinian tradition (including but not limited to the insights of neo-Darwinism) that is robust enough to explain both biological and cultural evolution without doing violence to either. We need an ecologically-grounded evolutionary theory that honors and protects individuals, communities and their traditions within a natural context, for we live in no other.

To return, then, to the question posed by the title, "Does the second law of thermodynamics refute the neo-Darwinian synthesis?" the answer depends upon what is meant by "refute." We have argued throughout this paper that the thermodynamical perspective does *not* render Darwinism or neo-Darwinism *false* in comparison to its evolutionary-developmentalist competitor. But, if the question is whether there is, or could be, a successor theory available within the Darwinian Research Tradition that avoids the conceptual and empirical disadvantages of the dominant neo-Darwinian sociobiology, then perhaps the answer is a qualified yes.

References

BOYD, R., RICHERSON, P.J.: *Culture and the Evolutionary Process*, Chicago (University of Chicago Press) 1985.

BRANDON, R.N.: *Adaptation and Environment*, Princeton, NJ (Princeton University Press) 1990.

BROOKS, D.R., WILEY, E.O.: *Evolution as Entropy: Toward a Unified Theory of Biology*, Chicago (University of Chicago Press) 1986.

BROOKS, D.R., WILEY, E.O.: *Evolution as Entropy: Toward a Unified Theory of Biology*, Second Edition, Chicago (University of Chicago Press) 1988.

DARWIN, C.R.: *On the Origin of Species by Means of Natural Selection, or the Preservation of Favored Races in the Struggle for Life* (Sixth Edition), London (John Murray) 1872.

DAWKINS, R.: *The Selfish Gene*, Oxford/New York (Oxford University Press) 1976.

DAWKINS, R.: *The Extended Phenotype: The Long Reach of the Gene*, Oxford/New York (Oxford University Press) 1982.

DENNETT, D.C.: *Darwin's Dangerous Idea: Evolution and the Meanings of Life,* New York (Simon and Schuster) 1995.

DEPEW, D.J., WEBER, B.H.: "Innovation and Tradition in Evolutionary Theory: An Interpretive Afterward", in: D.J. DEPEW, B.H. WEBER (Eds.): *Evolution at a Crossroads: The New Biology and the New Philosophy of Science*, Cambridge, MA (MIT Press) 1985, pp. 227-260.

DEPEW, D.J., WEBER, B.H.: "Consequences of Nonequilibrium Thermodynamics for the Darwinian Tradition", in: B.H. WEBER, D.J. DEPEW, J.D. SMITH (Eds.): *Entropy, Information, and Evolution: New Perspectives on Physical and Biological Evolution,* Cambridge, MA (MIT Press) 1988, pp. 317-354.

DEPEW, D.J., WEBER, B.H.: *Darwinism Evolving: Systems Dynamics and the Genealogy of Natural Selection*, Cambridge, MA (MIT Press) 1995.

DOBZHANSKY, TH.: *Genetics and the Origin of the Species*, New York (Columbia University Press) 1937.

DRAKE, J.A.: "Communities as Assembled Structures: Do Rules Govern Patterns", *Trends in Ecology and Evolution*, 5 (1990), pp. 159-163.

ELDREDGE, N.: *Unfinished Synthesis: Biological Hierarchies and Modern Evolutionary Thought*, New York (Oxford University Press) 1985.

ELDREDGE, N.: *Reinventing Darwin: The Great Debate at the High Table of Evolutionary Theory*, New York (John Wiley and Sons) 1995.

ELIOT, T.S.: "Tradition and the Individual Talent", in: *Selected Essays of T.S. Eliot*, New York (Harcourt, Brace and World) 1960, pp. 1-11.

FISHER, R.A.: *The Genetical Theory of Natural Selection*, Oxford (Oxford University Press) 1930.

GELL-MANN, M.: "Complex Adaptive Systems", in: H.J. Morowitz, J.L. Singer (Eds:): *The Mind, The Brain, and Complex Adaptive Systems*, Reading, MA (Addison-Wesley) 1995, pp. 11-23.

GOODWIN, B.: *How the Leopard Changes Its Spots: The Evolution of Complexity*, London (Wedenfeld and Nicolson) 1994.

GOODWIN, B., SAUNDERS, P. (Eds.): *Theoretical Biology: Epigenetic and Evolutionary Order from Complex Systems*, Edinburgh (Edinburgh University Press) 1989.

GOULD, S.J.: "Darwinism and the Expansion of Evolutionary Theory", *Science,* 216 (1982), pp. 380-387.

GOULD, S.J.: *Wonderful Life: The Burgess Shale and the Nature of History,* New York (Norton) 1989.

GOULD, S.J., LEWONTIN, R.C.: "The Spandrels of San Marco and the Panglossian Paradigm: A Critique of the Adaptationist Programme", *Proceedings of the Royal Society London B,* 205 (1979), pp. 581-598.

GRIFFITHS, P.E., GRAY, R.D.: "Developmental Systems and Evolutionary Explanation", *The Journal of Philosophy* , 91 (1994), pp. 277-304.

HO, M.-W.: "Environment and Heredity in Development and Evolution", in: M.-W. HO, P.T. SAUNDERS (Eds.): *Beyond Neo-Darwinism: An Introduction to the New Evolutionary Paradigm*, London (Academic Press) 1984.

HO, M.-W.: "On Not Holding Nature Still: Evolution by Process, Not by Consequence", in: M.-H. HO, S.W. FOX (Eds.): *Evolutionary Processes and Metaphors*, Chichester (John Wiley and Sons) 1988, pp. 117-144.

HO, M.-W.: *The Rainbow and the Worm: The Physics of Organisms*, Singapore (World Scientific) 1993.

HO, M.-W., SAUDNERS, P.T. (Eds.): *Beyond Neo-Darwinism: An Introduction to the New Evolutionary Paradigm,* London (Academic Press) 1984.

HODGE, M.J.S.: "Biology and Philosophy (Including Ideology): A Study of Fisher and Wright", in: S. SARKAR (Ed.): *The Founders of Evolutionary Genetics*, Dordrecht (Kluwer) 1992, pp. 231-293.

HULL, D.L.: *Science as a Process: AN Evolutionary Account of the Social and Conceptual Development of Science*, Chicago (University of Chicago Press) 1988.

KAUFFMAN, S.A.: *The Origins of Order: Self-Organization and Selection in Evolution*, New York, Oxford (Oxford University Press) 1993.

KAUFFMAN, S.A.: *At Home in the Universe: The Search for the Laws of Self- Organization and Complexity*, New York, /Oxford (Oxford University Press) 1995

LEWONTIN, R.C.: *The Genetic Basis of Evolutionary Change*, New York (Columbia University Press) 1974.

LEWONTIN, R.C.: "Facts and the Factitious in Natural Science", *Critical Inquiry*, 18 (1991), pp. 140-153.

MAY, R.M.: "Biological Populations with Nonoverlapping Generations: Stable Points, Stable Cycles and Chaos", *Science*, 186 (1974), pp. 645-647.

MAY, R.M.: "The Evolution of Ecological Systems", *Scientific American,* 239 (1978), pp. 160-176.

MAY, R.M.: "Chaos and the Dynamics of Biological Populations", in: M.V. BERRY, I.C. PERCIVAL, N.O. WEISS (Eds.): *Dynamical Chaos*, Princeton, NJ (Princeton University Press) 1987, pp. 27-43.

MAYNARD SMITH, J.: "Life at the Edge of Chaos?", *The New York Review of Books*, 42 (2) (1995), pp. 28-30.

MOROWITZ, H.: *Beginnings of Cellular Life: Metabolism Recapitulates Biogensis,* New Haven (Yale University Press) 1992.

ORNSTEIN, D.S.: "Ergodic Theory, Randomness, and 'Chaos'", *Science,* 242 (1989), pp. 182-187.

OYAMA, S.: *The Ontogeny of Information*, Cambridge (Cambridge University Press) 1985.

OYAMA, S.: "Stasis, Development and Heredity", in: M.-W. HO, S.W. FOX (Eds.): *Evolutionary Processes and Metaphors*, Chichester (John Wiley and Sons) 1988, pp. 255- 274.

OYAMA, S.: "Ontogeny and Phyologony: A Case of Metarecapitulation?", in: P. GRIFFITHS (Ed.): *Trees of Life*, Dordrecht (Kluwer Academic Publishers) 1992, pp. 211-239.

OYAMA, S.: "Constraints and Development", *Netherlands Journal of Zoology,* 43 (1993), pp. 6-16.

OYAMA, S.: "The Accidental Chordate: Contingency in Developmental Systems", *The South Atlantic Quarterly,* 94 (1995), pp. 509-526.

PRIGOGINE, I.: *Introduction to Thermodynamics of Irreversible Processes*, Springfield, IL (Charles C. Thomas) 1955.

SALTHE, S.N.: *Evolutionary Hierarchical Systems: Their Structure and Representation*, New York (Columbia University Press) 1985.

SALTHE, S.N.: *Development in Evolution: Complexity and Change in Biology*, Cambridge, MA (MIT Press) 1993.

SCHNEIDER, E.D.: "Thermodynamics, Ecological Succession, and Natural Selection: A Common Thread", in: B.H. WEBER, D.J. DEPEW, J.D. SMITH (Eds.): *Entropy, Information, and Evolution: New Perspectives on Physical and Biological Evolution*, Cambridge, MA (MIT Press) 1988, pp. 107-138.

SCHNEIDER, E.D., KAY, J.J.: "Order from Disorder: The Thermodynamics of Complexity in Biology", in: M.P. MURPHY, L.A.J. O'NEILL (Eds.): *What is Life? The Next Fifty Years: Speculation on the Future of Biology*, Cambridge, MA (Cambridge University Press) 1995, pp. 161-173.

SCHROEDINGER, E.: *What is Life? The Physical Aspect of the Living Cell,* Cambridge, MA (Cambridge University Press) 1944.

SEGAL, L., JACKSON, J.: "Dissipative Structure: an Explanation and an Ecological Example", *Journal of Theoretical Biology* , 37 (1972), pp. 545-559.

STRAUSS, S.Y.: "Indirect Effects in Community Ecology: Their Definition, Study and Importance", *Trends in Ecology and Evolution*, 6 (1991), pp. 206-210.

SWENSON, R.: "Emergent Attractors and the Law of Maximum Entropy Production: Foundations to a Theory of General Evolution", *Systems Research*, 6 (1989), pp. 187- 197.

SWENSON, R. (1991a): "End-Directed Physics and Evolutionary Ordering: Obviating the Problem of the Population of One", in: F. GEYER (Ed.): *The Cybernetics of Complex Systems: Self-Organization, Evolution and Social Change*, Salinas, CA (Intersystems) 1991, pp. 41-60.

SWENSON, R. (1991b): "Order, Evolution and Natural Law: Fundamental Relations in Complex Systems Theory", in: C. NEGOTIA (Ed.): *Handbook of Systems and Cybernetics*, New York (Marcel Decker) 1991.

SWENSON, R.: *Spontaneous Order, Evolution, and Natural Law: An Introduction to the Physical Basis for an Ecological Psychology*, Hillsdale, NJ (Lawrence Erlbaum Associates) 1997.

SWENSON, R., TURVEY, M.T.: "Thermodynamic Reasons for Perception-Action Cycles", *Ecological PSYCHOLOGY*, 3 (1991), PP. 317-348.

TURNER, J.R.G.: "Fisher's Evolutionary Faith and the Challenge of Mimicry", *Oxford Surveys in Evolutionary Biology,* 2 (1985), pp. 159-196.

ULANOWICZ, R.E.: *Growth and Development: Ecosystems Phenomenology,* New York (Springer) 1986.

ULANOWICZ, R.E.: "A Phenomenology of Evolving Networks", *Systems Research,* 6 (1989), pp. 209-217.

ULANOWICZ, R.E.: "Boundaries on the Complexity of Evolving Networks: A Window of Vitality", in: J. COLLIER (Ed.): *Between Instruction and Selection: Studies in Nonequilibrium Biology*, Baltimore(Johns Hopkins University Press) 1995.

WATSON, J.D., CRICK, F.C.: "Molecular Structure of Nucleic Acids: A Structure for Deoxyribose Nucleic Acid", *Nature* , 171 (1953), pp. 737-738.

WEBER, B.H.: "Complex Systems Dynamics and the Evolution of Biological Hierarchies", in: F. GEYER: *The Cybernetics of Complex Systems: Self-Organization, Evolution, Social Change,* Salinas, CA (Intersystems), 1991, pp. 31-40.

WEBER, B.H., DEPEW, D.J.: "Natural Selection and Self-Organization: Dynamical Models as Clues to a New Evolutionary Synthesis", *Biology and Philosophy,* 11 (1996), pp. 33-65.

WEBER, B.H., DEPEW, D.J., DYKE, C., SALTHE, S.N., SCHNEIDER, E.D., ULANOWICZ, WICKEN, J.S.: "Evolution in Thermodynamic Perspective: An Ecological Approach", *Biology and Philosophy,* 4 (1989), pp. 373-405.

WERREN, J.H.: "The Paternal Sex-Ratio Chromosome of Nasonia", *American Naturalist,* 137 (1991), pp. 392-402.

WICKEN, J.S.: *Evolution, Information and Thermodynamics: Extending The Darwinian Program*, New York/Oxford (Oxford University Press) 1987.

WILLIAMS, G.C.: *Adaptation and Natural Selection*, Princeton, Nj (Princeton University Press) 1966.

WILLIAMS, G.C.: *Natural Selection: Domains, Levels and Challenges*, New York, Oxford (Oxford University Press) 1992.

WILSON, D.S.: *The Natural Selection Of Populations And Communities,* Menlo Park (Benjamin Cummings) 1980.

WILSON, D.S.: "Complex Interactions in Metacommunities, with Implications for Biodiversity and Higher Levels of Selection", *Ecology*, 73 (1992), pp. 1984-2000.

WILSON, D.S.: "Variations on Themes of Darwinism Evolving", *Artificial Life*, 3 (1996), in press.

WILSON, D.S., KNOLLENBERG, W.G.: "Adaptive Indirect Effects: The Fitness of Burying Beetles with and without Their Phoretic Mites", *Evolutionary Ecology*, 1 (1987), pp. 139-159.

WILSON, D.S., SOBER, E.: "Reviving the Superorganism", *Journal of Theoretical Biology,* 136 (1989), pp. 337-356.

WILSON, D.S., SOBER, E.: "Re-Introducing Group Selection to the Human Behavioral Sciences", *Behavioral and Brain Sciences*, 17 (1994), 585-654.

WRIGHT, S.: "The Roles of Mutation, Inbreeding, Crossbreeding and Selection in Evolution", *Proceedings of the Sixth International Congress of Genetics*, 1 (1932), pp. 356-366.

Part II

Natural Selection and Developmental Systems: Redrawing the Boundaries Between Genetic and Non-Genetic Factors of Evolution

Chapter 4

Evolutionary and Developmental Formation: Politics of the Boundary

SUSAN OYAMA

I. Introduction

Much of my work has been concerned with what we could call the politics of the boundary. The meaning of "politics" here is very broad, having to do with all sorts of influence and power, but especially the power to define and privilege, include and exclude, render central or peripheral. Though this may involve matters "outside" science (a fraught frontier if ever there was one) it need not. Some of my reasons for working on the nature-nurture problem stem from concerns about publicly contested issues of, say, intelligence, race or sex, but most have to do with the kinds of distinctions that are made in the scientific work that draws on and feeds these larger disputes. Any theory carves the world in particular ways and so legitimates some entities and distinctions while leaving others beyond the pale--secondary, invisible or unintelligible.

Making the cellular or the nuclear membrane a primary theoretical boundary in the study of development and evolution, for example, may be justified by pointing to Weismann's barrier (1893), which forbids "information flow" from the body to the DNA. This is not just a matter of the presence or absence of cell-level DNA-altering feedback, however; it plays on and warrants a whole set of assumptions about inheritance and essence, permissible and impermissible explanations, even about academic disciplines (consider the distinction between biology and culture).

I am aware that, beginning with the words "politics" and "boundary", I am raising some red flags. Part of my project is to inquire into the provocative power of such boundary-marking emblems. They tend to imply a particular field of oppositions, and inevitably gloss over some differences and commonalities while proclaiming others, making it hard to move beyond global categorizations. This situation helps explain why there's a need for something like a developmental systems theory (DST for short), to rework many of the basic oppositions in the biological and social sciences. It may also help us understand why DST seems to be resisted for being simultaneously inside and outside of existing theory: inside (and redundant) because it stresses matters that are already acknowledged at least part of the time, and outside because it explicitly breaches certain sacred boundaries. (A fairly common reaction to DST is "That's a completely crazy idea, and besides, we already knew it.")

I start with a case for broadening the concepts of inheritance and evolution. Next come some recent border disputes in evolutionary theory. At various points I will touch on the selfish replicators that have figured in debates about "biological bases" of human behavior, and doubts will be raised about the status of "informational" units like genes and "memes" as privileged currencies and explanatory tools. More inclusive conceptions of development and evolution mitigate these difficulties and open up space to address issues that have suffered relative neglect. It may even be that they will allow the exploration of the complexities of cooperation and competition without foreclosing some possibilities beforehand. I close with some comments on the position of DST with respect to more mainstream theories, suggesting that some of the resistance it has met stems from treating certain boundaries as defenses against heresy, rather than as lines to be drawn and redrawn in the pursuit of various kinds of understanding.

I suggest that the boundary-shifting in developmental systems theory is not different in kind from the everyday work of science. This does not make it *right*, but it may make certain moves look less outlandish. DST is perhaps better known for denying distinctions than for drawing them, but repositioning a boundary can not only group together things that were formerly kept strictly apart, like genetic and other influences on development, but it can also throw into sharper relief differences that the usual geography minimizes, even as it brings the peripheral into focus. A much-analyzed case in point is the many meanings of innateness. The differences between presence at birth, for instance, and species-commonality, or between developmental fixity and heritability, may be acknowledged piecemeal from time to time, but they tend to disappear in the next study, paragraph, or even breath. The point is not that individual scientists are unable, or even unwilling, to make these distinctions (though they may be), but that the larger frame within which they work is such that the distinctions are unclear or unstable. The varieties of "innateness" tend to be treated as different manifestations of the *same* underlying reality, tied together by the belief that there is an internal essence or nature that accounts for them all (Bateson 1990; Gray 1992; Oyama 1985). These issues are central to this volume's theme of ethics, economics and sociobiology, and later we'll consider the possibility that the terms in which competition and cooperation are theorized may be similarly skewed.

My approach to these matters may look "deconstructionist" to some. I plead ignorance of the finer points of literary theory but suggest that if the label is apt, it is not because I'm claiming that all distinctions are equally good or that anything goes (Oyama 1995; B. H. Smith 1988 and in press). I do, however, take seriously the reasons for, and implications of, drawing lines in some ways rather than others, and so wish to attend to the way such decisions are supported. It may be that within the variation now to be found in more conventional theorizing there exist trends in the direction of something like developmental systems, despite the fact that they are not identified as such; in any case, closer attention to theorists' choices of measures and comparisons can illuminate the ways in which conceptual schemes are maintained and changed. This, too, is part of the politics of the boundary. Thus, by speaking of biology in this way, I do not intend to expose it to the world, unclothed and unlovely. But the emperor has *many* clothes, more than he usually wears, and certain ensembles, not yet tried, might serve him better.

II. The Bounded Gene

I start, then, with some observations on genetic inheritance and a brief introduction to the notion of a developmental system. The conceptual line around the genes is fundamental to modern biology, and insofar as biological science plays a foundational role *vis a vis* other disciplines, other lines are drawn in accordance with it. It delimits inheritance proper, the genetic inheritance that was originally modelled on the inheritance of property or titles, but now serves as the paradigm case. The contrast class to genetic inheritance may be the phenotypic, the acquired, the developed, the learned, the cultural--one of the points of the critical literature on the nature-nurture dichotomy is that these are not the same contrasts, and that, even when they are distinguished, they are often ill-formed (Bateson 1991; Gray 1992; Hinde 1968; Ingold 1995; Johnston 1987, 1995; Lehrman 1970; Oyama 1985; Wimsatt 1986). A semipermeable membrane around a certain kind of cell, or even around its nucleus, is thought to separate two kinds of "information", "transmitted" through different channels. Genetic information flows along the germ line; it passes through organisms without being (much) altered by the passage and continues for an indefinite number of generations. Hence the talk of potentially "immortal" genes (Williams 1974, p. 24). The ontogenies of individual organisms are said to result from the decoding of this information, which may then be supplemented or modulated by the environment.

Once a line is drawn between the genes and everything else in the universe, it is perhaps not surprising that the channel for *non*genetic information remains underspecified (it may be cytoplasmic or extracellular, "experiential" or cultural), though a direct brain-to-brain route seems to be popular. This story about information involves a restricted (and restrictive) logic. Evolution is narrowly conceived as change in gene frequencies. Heredity is narrowly conceived as gene transmission. Only genes pass through the bottleneck of reproduction, so the next generation must be explicable in terms of those genes (Raff/Kaufman 1983, p. 234). Development is blackboxed, ultimately attributed to gene action. Intriguingly, despite this blackboxing of the processes of developmental formation, contemporary evolutionary treatments both assume and legitimate a particular view of development that I have called "developmental dualism", the doctrine that some developmental processes are (mostly) guided from within, by the genes, and some are (mostly) guided from the outside. This is so even though evolutionary and developmental theorists alike frequently declare the nature-nurture distinction to be obsolete.

Sober (1992) gives a taxonomy of selective models of cultural change based on the use of either genes or learning as the mode of heredity and either "having babies" or "having students" as the measure of fitness. We see here both the centrality of information-transmission as a mechanism of continuity and the intimate connection of these models with the nature-nurture divide. It becomes a simple matter to see how this conceptual geography encourages genetically reductive views of human affairs (if only gene replication counts) and/or developmental dualism (especially if "memes", or cultural units, replicate too).

III. Redrawing Boundaries

Developmental systems theorists have proposed that the overly restrictive boundary around the process of heredity be redrawn to include other developmentally important influences (Ho 1988 gives an overlapping treatment). The point is not to open a second channel, to carry culture, but to trade discrete channels for interacting systems whose processes give rise to successive generations. Regularities in these systems' functioning support the sorts of predictable associations that make the language of transmission possible.

1. Assuming Reliable Development

Transmission, which masquerades as a mechanism explaining the regularity of development, actually *presupposes* and *requires* reliable development. For a given DNA sequence or phenotypic feature to reappear in the next generation, both parent and offspring must have developed well enough to be viable and similar in the relevant ways. "Cultural transmission" also assumes a great deal. When beliefs or behavior patterns recur regularly we can count on them without being able to explain them. When they are unpredictable we are frequently at a loss to say why, though there is no end to speculations. Think of everything that must be in place for generations in a family to become fiscal conservatives, again and again, or Baptists (or Darwinians). The fact is, we *can't* think of everything that would be required, because we don't know.

Such phenomena are notoriously hard to explain; an account of imitation or family dynamics certainly won't suffice, and neither will outside-in explanations in terms of class or culture. How often do we accurately transmit even a single complex idea to a student, or a colleague? (What would it mean to transmit a "single idea" in isolation?) Something is judged to have been transmitted when it reappears, through heaven knows what processes, in the "recipient." Notice, though, that invoking transmission not only fails to *explain* how this developmental/cognitive reconstruction occurs, it encourages us to ignore the fact that it must occur at all.

In a discussion of cultural evolution, Sober (1992) says social scientists are interested in *causes* of differential transmission, and evolutionists, in *consequences.* (A version of this perspectival gap will be seen again, when we talk about cooperation and competition.) Thus Dawkins (1976, p. 206) can blithely write of memes that "propagate themselves in the meme pool by leaping from brain to brain" precisely because, as Dennett (1995, p. 359) acknowleges, evolutionists and cognitive scientists can "finesse" their "ignorance of the gory mechanical details of how the information got from A to B." One could suspect that more than finessing is involved here, given the evident glee with which they displace agency from humans to selfish replicators. Thus Dennett speaks of "invasion of human brains by culture" (1995, p. 369), and even of interacting "meme-infested brains" (not people!) (p. 471). True, in a more benign mood he says brains are "furnished by" memes (p. 341) but he actually *defines* the notion of a person by such memic invasion and manipulation. The general thrust is conveyed by his description of selves as "created out of the interplay of memes exploiting and redirecting the machinery Mother Nature has given us" (p. 367).

These are just especially vivid examples of the ways in which not only developmental issues, but all sorts of questions about social behavior, are kept at bay (or worse, "explained") by evolutionary theorists. If evolutionists are content to *assume* the processes allowing reliable reappearance across generations, and wish only to document net outcomes, then one might reasonably ask what purpose is served by the agentic language, inviting as it does precisely the conflation of process and consequence that Bateson (1988) warns against, especially when it draws so much critical fire. Yet self-interested replicators go on jockeying for reproductive advantage. They are prototypes of competitive selfishness, and they design us to be their instruments (see critiques in Bateson 1986; Kitcher 1985; Oyama 1985, 1991a). Small wonder that this style of evolutionary writing has given rise to worries about the possibility, even the coherence, of cooperation, as well as about the more general implications of evolution for human lives.

2. Privileging Genes

In standard accounts of evolution and development, genes are privileged, as both currency and cause. DST contends that the privileging is unjustified. Multi-levelled, changing contexts and processes can only be reduced to the role of conduits for information by ignoring the "gory mechanical details" of the life cycle. To give a flavor of developmental systems-style revisionism, here are five typical rationales for gene-based accounts, each with a parenthesis indicating some of what is being glossed over. There is nothing sinister about these glossings-over: We simplify all the time. Nor is there anything particularly heterodox about them. In fact, they are phrased in a conservative manner, as I think the kind of speaker I am imagining would qualify the arguments if it became necessary. Parentheses can, however, marginalize by typographical convention, minimizing what van der Weele (1995) calls the "conceptual room" allocated to a topic. She speaks of scientific choices in terms of the "ethics of attention"; one can similarly speak of a politics of attention. So each parenthesis is expanded, to place its contents in a developmental systems frame in which the justification of the special status of genes is applied, by parity of reasoning, to other developmental influences as well. In DST, all of these influences and entities, both inside and outside the organism, are interactants in a developmental system that produces a life cycle. Herewith the arguments, with parentheses and their expansions:

a. Genes produce organisms. (Though they are not, of course, sufficient; materials must be available and conditions must be adequate.)
   EXPANSION: Genes themselves don't "make" anything, though they are involved in processes requiring many other molecules and conditions (Moss 1992, in prep; Nijhout 1990). Additional interactants (or resources, or means) are found at scales from the microscopic to the ecological, some living, some not. None is sufficient and their effects are interdependent. Development never occurs (and could not occur) in a vacuum.
b. Shared genes are responsible for species characteristics. (Though again, they are not sufficient. Proper conditions must be present and it's true that genes are involved in all characteristics, species-typical or not.)

EXPANSION: Just as genes can't make organisms *in general*, they can't create species-typical characters *in particular*. Typical conditions, again at many scales, contribute to forming these characters, the uniformity of which should not be exaggerated. The activity of the organism, including self-stimulation, is often a crucial aspect of species-typical development (Gottlieb 1978; Johnston/Gottlieb 1990; Lehrman 1970; Oyama 1982). Both genetic and environmental variation are often underestimated, and flexible processes can sometimes result in typical phenotypes despite atypical developmental resources.

c. Genetic variants specify the heritable phenotypic variants necessary for natural selection. (Though heritability does depend on conditions and on the particular comparison being made, and it's often hard to separate genetic and environmental effects.)
EXPANSION: Unless nongenetic factors are excluded by stipulation, other developmental resources can also "specify" phenotypic variants, which can be heritable in a variety of senses. The genotype-phenotype correlations that warrant the talk of genetic specification may not occur under all circumstances, and may change. Specificity is a slippery matter; it depends on the question being asked, the comparison being made, developmental state of the organisms, and the context of the comparison, as well as the level of analysis. In fact the genotype-environment correlations and statistical interactions that plague the population geneticist are manifestations of the interdependent networks that developmental systems theorists describe (Moss, in prep.; K. C. Smith 1994).

d. Only genes are passed on in reproduction; phenotypes, and therefore environmental effects, are evanescent and thus evolutionarily irrelevant. (Though the genes are housed in a cell.)
EXPANSION: If transmitting or "passing on" means "delivering materially unchanged", then few if any developmental resources are transmitted across evolutionary time (depending on how you measure material change). If transmission means "reliably present in the next life cycle", which is the biologically relevant meaning in DST, then an indefinitely large set of heterogeneous resources or means is transmitted, sought or produced by the organism itself, supplied by conspecifics or others, possibly through social processes and institutions (Caporael 1995; Ingold 1995). Although many developmentally important environmental features are exceedingly stable, others are noncontinuous,

perhaps varying seasonally or geographically (Griffiths/Gray 1994). Any definition of inheritance that does not privilege the nuclear or cell boundary *a priori* will be applicable to other constituents of the system. The developmental systems perspective stresses the processes that bring together the prerequisites for successive iterations of a life cycle. I have observed elsewhere (1988, p. 271, note 6) that bits of DNA could magically materialize whenever they were to be used, that it is appearance at the right time and place that counts, not material continuity. The more we learn about how cells carry on, the less fanciful this seems; a functional gene may not consist of a continuous stretch of DNA but may have to be assembled on the spot by quite elaborate methods (Moss 1992, in prep; Neumann-Held, this volume, in press).

e. If gene frequencies don't change, then evolution has not, by definition, occurred. (Of course, the gene concept is relatively recent, and other definitions are possible.)
EXPANSION: An historian could tell us how gene frequencies moved from being an *index* of evolutionary change to being *definitional*, but we needn't insist on that one definition. In fact, many branches of biology routinely speak of changes in phenotypes (Johnston/Gottlieb 1990; Maynard Smith 1978/1984 on game theory). If one must have a "unit" of evolution, it would be the interactive developmental system, which includes the organism but is emphatically not restricted to it. Evolution would then be change in the constitution and distribution of these systems. This definition allows, but is not restricted to, more traditional ones.

IV. Negotiations at the Border

This should give a sense of the DST approach to development and evolution. We extend some boundaries, but in many ways we are simply making explicit what is everywhere implicit, and what increasing numbers of workers are saying more or less directly. In another, more realistic sense, however, it looks like an uphill fight all the way. That these are both true is in itself important, for it reveals the complexity of the process of theoretical conservatism and change (Shanahan, in press; B. H. Smith, in press).

Brandon and Antonovics (1995, p. 229), for instance, claim that organisms and their environments coevolve. They discuss Richard Lewontin's (1983) argument that organisms and environments are interconnected in ways that require a reconceptualization of evolution. They note that the idea of such mutual influence has been around a long time (in the practice of rotating crops, for example). It is also in the evolutionary literature, often in asides and *ceteris paribus* clauses, but prominently in game theory, which may be easier to accommodate because it involves competing organisms. The commentator to Brandon and Antonovics' article (Wolters 1995) is appreciative, but disagrees with them on whether conspecifics are internal or external to the population. Border disputes like these are everyday affairs, and such uncertainties, like the longstanding one about whether the niche belongs to organisms or the environment, illustrate both the perpetual ambiguity of these divides and the felt need of scientists to bridge them.

1. Using What You "Know"

Theorists are exasperated to be told what they have "always known." Yet there is a difference between knowing in a parenthetical, "of course it's important" way about the intimacy and reciprocality of organism-environment exchanges in development and evolution, say, and incorporating the knowledge in models and explanations, research and theory. Indeed, one of Brandon and Antonovics' stated goals is to redress evolutionary biology's failure to integrate this "knowledge" of interdependence. Cor van der Weele (1995), remarking on the same phenomenon in developmental biology, shows some of the ways the role of the environment is marginalized. Because any dividing line exists on a complicated, multidimensional landscape of belief and practice, it can implicate all sorts of other affirmations and denials, to say nothing of loyalties and betrayals. A great deal of energy is devoted to managing them.

2. Moving the Line

Negotiations over entities and explanations are integral to the making of science, including the continuous "making" of science by marking the borders between it and nonscience. Whenever the scope of a theory is being explored or unexpected results must be accommodated, limits are put into question. A common response to theoretical challenge is to engage in strategic boundary work: Hull (1988, p. 202) holds that the proponents of the evolutionary synthesis simply redefined the theory to include neutralism, for instance, and Godfrey-Smith (1996) supplies more examples of creative line-drawing. As suggested earlier, one can also attempt to place an opponent beyond the pale, perhaps by calling him or her a Lamarckian. There is an interesting story to be told about this use of Lamarck against any deviation from traditional conceptions of development and evolution, but I will not try to tell it here (for comments, see Hull 1988, ch. 12, and Sterelny/Smith/Dickison 1996). Suffice it to say that it is often a dubious attempt to reinforce a dubious distinction between biology and learning, nature and nurture, between hardheaded, scientific neoDarwinism and fuzzyheaded sentimental mysticism. True, cultural change may be termed "Lamarckian", but biology itself must be shielded by Weismann's barrier or some other line between the genetic and environmental, despite the difficulty of saying just where it lies and despite the myriad intimate connections and interdependencies between the divided territories (Latour 1993).

We have seen that variously located membranes are used simultaneously to demarcate kinds of information, inheritance, transmission, evolution, even science itself. (See Sapp 1987, on battles over the nucleocytoplasmic gap in early genetics.) Most of the lines touched upon here have been, as they say, internal to science, but they are political nonetheless, aspects of disciplinary and doctrinal influence. The skeins of citations (as well as the lacunae left when critics are answered or dismissed without being named) are also boundary work, creating ancestors, bastards, ingroups and outgroups: work that is *at once* both intellectual and political.

The runoff from these controversies is seldom contained by ivy-covered walls. Debates about macroevolution or developmental constraints, for instance, have become ensnared in the battle between creation science and orthodox science. Part of the emotional charge to these debates comes from exasperation that a battle considered long won is not over after all. Theological considerations were once standard in biological inference. That changed, and some would change things back; theorists accustomed to talking about escalating arms races should expect to face updated weapons. As Hull (1988, p. 73) remarks, "The dispute over Darwinism was as much a disagreement over the nature of genuine science as over the existence of evolution." It still is.

V. Replicators, Cooperation and Competition

Once one has accepted a gene-based view of biological evolution one can then ask: If genes evolve in particular ways, why not other things? If tiny particles of matter are central to one story, atoms of thought or culture could populate another (note the way the body-mind duality is built into the nature-nurture distinction). Many approaches to learning, society and culture have been inspired by evolution. Sometimes social arrangements are somehow explained by biology, or the relation between biological and cultural evolution may be analogical: they may "interact" or compete, or run in parallel. We might then envision gene-meme (or culture) coevolution. Dawkins (1976) created the unit of cultural inheritance, the "meme," on the model of his selfish gene. Just as genes could manipulate a bird into making a nest to ensure their propagation (Dawkins 1983), an idea or tune could arrange for its own replication by imitation (hence the term, replicator). There is some parity of reasoning going on here, and scheming memes can be as entertaining as selfish genes. But as Bateson (1978) suggested in a playful but pointed rejoinder to Dawkins, one could also speak of nests using genes to make more nests.

1. A Proliferation of Replicators

Serious proposals to enlarge the evolutionary cast of characters tend to be resisted. One of the most conservative elements in standard views is a belief in entities whose "informational" nature gives them special significance in both evolutionary and developmental formation. "Conservative" here refers to theory, but one of the reasons evolutionary theory demands so much attention from scientists and nonscientists alike is its entanglement in broader political and moral questions. Having to announce "I am not a Nazi" or "I am not a lackey of the capitalist power structure" is about as much fun as going on television to say "I am not a crook." Scientists offended at having to parry such accusations may, however, have helped prepare the way for them by cleaving to a conceptual scheme in which human relations are explained by the competition of immortal quasi-agents that get themselves counted in the next generation by making and running the the bodies that carry them. Or less melodramatically, a scheme in which a biologically conservative "nature" constrains and limits "nurture", defining beforehand the range of possible variation. (Not that one can avoid being called a Nazi by taking a critical stance toward these ideas; culture theorists and relativists of every stripe hear the charge, too. As suggested above, those who question received evolutionary wisdom now risk being labelled creationists as well. The point here is not that one can totally control the names one is called, but that it is sometimes possible to gain some insight into the ways name-calling works.)

The lively and ongoing exchange between Sterelny, Smith and Dickison (1996), and Griffiths and Gray (1994/in prep.) is, among other things, a second-generation Dawkins-Bateson disagreement about the notion of the selfish replicator, which for Dawkins (1982) (Hull 1988 has a different view) is an entity that makes organisms and is copied at differing rates, depending on the organisms' reproductive success (see Griesemer, in press), on copying). Sterelny, Smith and Dickison continue to insist on the privileged *evolutionary* status of the replicator, though they accept "the radicals'" critique of its special role in controlling *development.* (For "radicals" read "DST." This privileged role in developmental causation is seen in Dawkins and many others in the guise of genetic programs and the like.) The "reformists" (my term, not theirs) not only wish to *retain* evolutionary replicators, they want to enlarge the category, to include more entities: "a still more raucous and motley crowd of squabbling replicators", like nests and burrows. They note that treating such objects as replicators is consistent with Dawkins' "basic conceptual structure", if not with his "actual practice" (see also Gray 1992). But DSTers Griffiths and Gray question the very idea of privileged replicators. The choice, to simplify somewhat, is between reforming the game, by admitting more players, and changing it. This game is enclosed in a bigger game that is not being questioned, maybe the game of "doing evolutionary theory" or "scholarly exchange."

This complex debate cannot be summarized here, but it is interesting partly because evolutionary theory, which eliminated divine agency from its explanation of life, can end up postulating surrogate agents. This metaphorical talk is not entirely benign, and the difficulties are not completely defused by accusing critics of squeamishness about real life, or by invoking poetic license. One way it spawns trouble is by making the already difficult task of thinking about human behavior even more difficult. In most cases this is probably an unintended consequence. Dennett's (1995) autonomous, hyperactive genes and memes, however, seem rhetorical aids to his denial that humans have original intentionality; more generally, selfish-gene-talk is an aspect of an evolutionary *machismo* directed against anyone foolish enough to think that nature (or humanity) is nice.

Though I don't think nature is particularly nice, sometimes people are, and their niceness is not always a selfish strategy. I am concerned about the kinds of truths evolution is thought to render, and believe the problems raised by current orthodoxy's genecentrism go beyond the theoretical issues sketched earlier. The risk that we will see humans as being driven by the self-interested replicators that "infest" their bodies and brains is somewhat diluted if burrows and nests can be replicators, too. But crowds of contentiously quarreling quasi-agents may not be the best basis for an adequate view of human life. We are left with more competitors, not a different view of competition.

2. Evolving Systems

If one adopts DST's view of evolution as change in the constitution and distribution of developmental systems, it is certainly reasonable to record the reappearance in successive generations of particular features, variant or otherwise, as a way of tracking the characteristics of life cycles. Historical considerations, ease of measurement and amenability to modeling will inform the choice of features and comparisons. Stability and change in associations among features can also be of interest (this is what studies of heritability, linkage, habitat imprinting, social structure and migration are about). Virtually all the familiar kinds of evolutionary research can still be done against such a broadened background, but no elements are privileged *a priori*, either informationally or causally. There is, though, special emphasis on the many explicit and tacit choices made by the analyst who temporarily focuses on certain elements as "informative" (Moss 1995).

Thinking in terms of evolving systems (rather than disembodied gene pools, or genetic programs) makes the kinds of complex relationships alluded to by Brandon and Antonovics (1995) salient rather than practically invisible, as these authors argue has hitherto been the case. Their term, "coevolution", though, suggests the linked change in two distinct units brought about by natural selection of each by the other (typically different species, as in the host-parasite models the authors say are closest to their own, p. 227). I am not arguing for any single definition of coevolution (see Nitecki 1983), but I suspect that, to the extent that Brandon and Antonovics treat their plants and environments as somehow analogous to two species, they fall short of effectively serving their stated goal of implementing Lewontin's insights into organism-environment interpenetration, and to the extent that they approach that goal, they also approach something that looks more and more like a developmental system.

Griffiths and Gray (in prep.) compare the organism-environment coevolution of Brandon and Antonovics (1995) with organism-environment relations in evolving developmental systems. They address the supposedly intractable complexity of developmental systems and the necessarily limited scope of any actual investigation, remarking on the gap between the rhetorical force of a *charge* of wooly holism (levelled by Sterelny/Smith/Dickison 1996) and its *practical significance*. They point out that a "complete" picture of Brandon and Antonovics' organism-environment coevolution would be as complex as a "complete" developmental systems account, but that any real analysis is circumscribed. (On the goal of completeness see van der Weele 1995)

Earlier I offered some speculation about sources of resistance to the pleasing (to me) unity and inclusiveness of the developmental systems account. But one need not haul out poor Lamarck in order to complain that speaking of evolving organism-environment systems involves an unacceptably great change from existing usage. Even Brandon and Antonovics' largely sympathetic commentator (Wolters 1995) feels that because "evolution" and "coevolution" have traditionally referred to genes and organisms it's a bit much to talk about the *environment* evolving. (Having applied the idea of replicator to extraorganismic entities, Sterelny/Smith/Dickison (1996) do not balk here; this is a consequence of taking an idea more seriously than its author does.) Apparently defining evolution by selection, Wolters also seems to say that environments are not naturally selected, so they *change* but do not *evolve* (see Moss, in prep., and Weber/ Depew, this volume, for some reworking of selection).

But Gray (1988) points out that Darwinism has no timeless essence. "Existing usage" is seldom univocal; common terms often hide decidedly varied meanings, and over time theories can be marvelously elastic. At any moment, the amount of heterogeneity of opinion within a tradition can be stressed or minimized; when threatened from "the outside" one circles the wagons rather than detailing internal disputes. Hull (1988, p. 200) for one has opined that "In retrospect, Huxley's appellation ["Synthetic Theory"] was hardly based on past accomplishments but was a combination of a public-relations ploy and a hope for the future" and *Keywords in Evolutionary Biology* (Keller/Lloyd 1992) shows even fundamental concepts like "gene", "fitness", "selection" and "species" are anything but monolithic. This does not keep them from being used, and it doesn't necessarily hinder discussion. Heterogeneity can feed flexibility and extension (Fujimura 1992). Developmental systems theorists (dis)respectfully submit additional concepts, like "information", "transmission", "inheritance", and "evolution" itself for scrutiny and reevaluation.

3. Indicators and Interests

There are perennial worries that we as a species are necessarily warlike, or patriarchal, and counterclaims are made that we are basically peaceful, or matriarchal. To speak of biology is often to speak of what is essential, deep and really real, so it is not surprising that people feel the need to fight evolutionary fire with more (albeit nicer) evolutionary fire, perhaps countering innate selfishness with innate altruism (Kohn 1990; comments in Oyama 1989, 1991b). This is unnecessary, and in the long run, counterproductive. Nor is the solution to say we have no nature, for this is apt to be taken as a claim that we can be anything, or else as a denial of within-species commonalities. To avoid these traps, I have recommended that "nature" simply refer to the organism's characteristics--to the phenotype throughout the life cycle. Natures, then, change, and organisms with the same genotype can have different "natures." "Nurture" then refers to the developmental processes that continuously produce and alter these natures. This removes the opposition between things that are read out from genetic information and things that are shaped from the outside.

Earlier I referred to the a widespread belief in a single reality underlying the various meanings of innateness, suggesting that the case of competition might be in some respects comparable. There is more to the comparison, because what concerns some people is precisely the idea that selfishness and competitiveness are part of our biological nature (in the usual senses, as fundamental or inescapable). Just as it is hard to avoid the idea of a unitary nature, it has become hard to think of organisms, including ourselves, as anything but competitive. This is not only due to images of striving replicators; assuming cooperation to be, at best, a competitive strategy can make it conceptually unstable. Keller (1992) observes that while mutualisms and cooperative relationships are acknowleged in biology, they tend to be subordinated to competitive ones, at times even being called "cooperative competition."

Competition is seen to be *caused* by limited resources or is simply *defined* by them. Resources are assumed to be measurable without reference to organisms, she reports, and an organism's consumption is similarly treated independently of other organisms. All this contributes to the virtual unthinkability of cooperative relationships except as subservient to competition. She further remarks on the apparent inability of the field to retain certain periodically-rediscovered insights about the dynamics of cooperation. This looks like insufficient conceptual room, or, as she puts it (1992, p. 120), "air space."

The literature is full of strategic altruism and mutual back-scratching, but always with an eye to reproductive advantage. Why must we keep *our* eye on reproductive advantage, itself a multivocal notion (Keller/Lloyd 1992)? Because, we're told, natural selection's invisible hand is guided by an eye fixed on the bottom line. But why just one bottom line?

Keller reports that (1992) evolutionary theorists speak of competition in the absence of direct conflict or contact, and even when the organisms are not using the same resource: what makes the relation competitive is that the *analyst* is making a comparison. "This extension, where 'competition' can cover all possible circumstances of relative viability and reproductivity, brings with it, then, the tendency to equate competition with natural selection itself" (p. 125). To continue the conceptual slide, remember that some people virtually identify evolution with natural selection, which is defined by genetic change. Taylor (personal communication) confirms that researchers often do not actually observe the use of a limited resource by different species, and that multiple indirect effects confound the inverse relations between population sizes from which ecologists have typically inferred competition.

Let's reflect a moment on genetic advantage. Dennett (1995, pp. 327-9) defends the idea that genes can have interests by comparing them to any other entity for whose benefit things are done, like children, corporations and ideas. Natural selection is directed at advancing the welfare of genes, he says, so genes have interests. Though he stops a hair short of flatly denying that all this is basically bookkeeping (Wimsatt 1980/1984) he makes it quite clear that he thinks these "interests" are not just a matter of what the biologist is counting: Interests are conferred by Dame Selection, not the scientist.

Though this account seems odd, one can certainly track allele frequencies from generation to generation, *if this is what you want to know.* Similarly, one can count the noses of a motley of replicators, but one should not forget who is choosing to count, and one should avoid straitening the conceptual space by adopting overly restrictive definitions or confusing the interests of the countee with those of the counter. The dominance of population-genetic definitions was established with the synthesis, which has its share of gaps, loose ends and indeterminacies. Without being so audacious as to present "better" definitions of cooperation and competition, I want to speculate a bit about collective activity.

4. Developmental Systems-Style "Competition" and "Cooperation"

I do not want to cast "the radicals" as exclusive press agents for cooperation. As noted above, evolutionists have studied it intensively, and Griffiths and Gray (1994) stress the importance of competition (as dependence on limited shared resources) in DST. The idea of competition is considerably altered, though, when focus shifts from organisms or agent-like replicators to systems of interrelated processes (see Griffiths/Gray, in prep.). At issue is the evolutionary *role* given to the complex interdependencies and integration highlighted in DST. What follow are some tentative comments about these ideas in what might loosely be called a developmental systems style: one that moves with a certain fluidity among scales and measures, takes a pragmatic stance toward research decisions and a somewhat skeptical one toward much received wisdom. This is a background theory that emphasizes emergent pattern from shifting, heterogeneous sets of interactants, and changing, multiple control (see also Taylor 1995). Such phenomena, which some call "cooperative" are frequently described by ecologists, developmentalists, and social scientists. They tend to appear in evolutionary theory only as constraints on, or conditions for, selection, as messy details to be bracketed out, or as strategies to propagate genes. Though present in the theory, then, they are rendered conceptually secondary. There is no such prior privileging in DST. These aspects of biological processes can be approached in a way that does not always measure them against the same bottom line. One can concentrate on *consequences* of variant developmental systems' differing rates of self-perpetuation or on the interrelated *causes* involved in the systems themselves. In the latter case, the "bottom line" is the continuity of a system's functioning, whether reliably repeated or not.

In "The Accidental Chordate" (1995) I spoke of the importance of understanding how a set of processes *keeps going*, partly by recreating its own constituents (NB: *not* how some subset of constituents recreates itself, making and using the rest of the system to further its own proliferative interests). I was speaking of autocatalytic chemical reactions in vital processes, but the significance of interacting, self-perpetuating and self-preparing complexes of resources and processes is more general. Larger-scale processes are implicated in reproduction and ontogeny as well, and they are not necessarily contained within a skin (one of the points of Dawkins' extended phenotype, see Dawkins 1983). The continuity of the germ line is supposed to explain the repeated cycling of life courses, but this leaves out the rest of the developmental complex, downgrading important phenomena and encouraging problematic conceptions of all-knowing creator-genes. Disclaimers of complete genetic efficacy are becoming more common, but they tend to leave the basic message of all-powerful macromolecules largely untouched. If there is indeed a practical difference between merely acknowledging phenomena and giving them full-fledged theoretical status, then it could be worthwhile to look at evolution through different lenses.

If we ask how systems keep going, how they change or remain stable, how changes at one level are or are not reflected at others, there will be room for tracking particular units from generation to generation, though they will not have unique status *a priori,* so analysts may be called upon to justify their decisions. A benefit of refocusing from tightly bounded self-replicators to loosely bounded repeating systems, however, might be to allow a more discriminating and generous view of collective activity (Gordon, in press). Just as development has never been fully integrated into the synthetic theory, Taylor (1992) observes that the question of long-term stability of multispecies communities has not been adequately dealt with. He stresses the historicity of ecological complexes, which must be developed or reconstructed, not just "dispersed" (see also Gray 1988).

Faced with a complex system made of processes at scales from the molecular to the biogeographical it is necessary to ask: Cooperation *for what?* Competition *for what?* There need not be just one bottom line (B. H. Smith 1988, ch. 6), and for any indicator to track *anything* the system must be kept going. Processes and entities are often at least partially nested, and there may be different consequences of an interaction at different levels (Wimsatt 1980/1984). Hence the earlier allusion to competing theorists "cooperating" to sustain larger games (which themselves may be in higher-level competition). At any level the entities or processes must be scrupulously specified, as must the relevant outcomes. If an analyst invokes competition when outcomes are *compared* (recall Keller's point), the counterpart could be to speak of cooperation any time an interaction contributes to the *creation* or *maintenance* of the outcomes themselves (Griffiths/Gray, in prep.), though using loaded terms like competition and cooperation for such broad categories of consequences and causes is probably ill-advised. In any case, these seem to different kinds of questions, such that either can be subsumed by the other by embedding. If the highest-level game is about relative *representation* in an outcome, then "cooperative competition" could make sense. If it's about *contributions* to an outcome, then "competitive cooperation" becomes intelligible. For both kinds of questions a variety of indicators and currencies could be considered, but one is not *automatically* subordinated to the other; which is primary depends on the analytic question. It would then become difficult to conclude anything about the "basic" nature of behavior.

Taking this kind of multilayered approach to human interactions reveals social behavior the variety, shifting subtlety and complexity of which defies straightforward categorization. Think of Goffman's (1959) classic analysis of self-presentation. People manage these presentations in ways that may be viewed as competitive and at the same time enter into wordless collusion to preserve a certain definition of the situation. The mobile psychology of "us" and "them" can also be of interest here (Caporael/Dawes/Orbell/van de Kragt 1989); think of the people/things you might be proud of, defensive of, embarrassed for, and the conditions under which those propensities can be aroused and altered, sometimes moment-to-moment. Trust is also crucial in our collective action, including science (Bateson 1988; Shapin 1994).

VI. Locating Developmental Systems

I am not advocating that we abandon existing methods of studying competition and cooperation, but that we take very seriously their limitations, and then ask, is there anything else, perhaps quite different, that we might want to know, as scientists or as citizens? The shape of the theoretical background surely influences the ease with which, and the ways in which, other questions come to mind, as well as our interpretive resources (Wimsatt 1980/1984).

Newspapers tell us that a single product can come to dominate a market for reasons other than absolute superiority. The essays in *Keywords in Evo-*

lutionary Biology (1992) are instructive, not only for the contemporaneous and historical variety of definitions, methods and assumptions surrounding important concepts, but for the stories that can be read between the lines and entries, about the dyamics of a sometimes loosely associated set of developing research areas, impinging upon each other, requiring coordinating concepts, flourishing or not (Callon 1991; Taylor 1995). A historical perspective can also help us understand how topics like cooperation and environmental influences in evolution (other than selective ones) gain the power to contaminate, in these cases by association with group selection and Lamarck (Shanahan, in press).

When I speak of a politics of theoretical boundaries, I do not mean that people are always maneuvering to push some views and suppress others, though they often are. But unintended consequences are ubiquitous. Convenient measures tend to turn into definitions, and not only in science (think of IQ, or grading in school), despite individuals' explicit desires that this not happen. (Psychologists are taught that Binet wanted to avoid precisely the use to which intelligence tests were eventually put.) Specialized measuring technologies arise, squeezing out other perspectives as intellectual and professional interests become linked in complex networks. Certain kinds of analyses and explanations are readily available, while others are hard even to think about. A colleague confides, "We don't have the language" to describe social behavior.

Though developmental systems treatments have been criticized for being too radical, when specific points are confronted, the response is often that they have always been acknowledged within conventional theory. I would not want the potential usefulness of this approach to be missed because topics like cooperation (not assessed in an exclusively competitive frame) and environmental influences (not treated exclusively from the point of view of selection) have been stigmatized. Classification as radical or moderate implies a single point from which distance can be measured, but I don't think that evolutionary theory has such a center (though natural selection comes close). Nor would I want valid criticism to be hastily brushed off with a reflex "We knew that", when what is needed is something more. Some of what we say and do is indeed acknowledged by most evolutionists, but you don't have to be a developmental systems theorist to see the difference between some familiar-looking bits and a novel but consistent configuration. In one analysis (van der Weele 1995) DST has similarities to, and differences from, both standard neoDarwinism and process structuralism. Clarifying these and other relationships (Oyama, in press) may help bypass some reflex objections and answer others, but the judgment about whether work in developmental systems lies inside or outside the synthetic theory can probably be made only retrospectively. Even then there is no guarantee that the judgement will be either unanimous or enduring.

ACKNOWLEDGMENTS: I am grateful for comments on this esssay by Pat Bateson, Linnda Caporael, Russell Gray, Paul Griffiths, Lenny Moss, Barbara Smith, Peter Taylor and Rasmus Winther. Earlier, shorter, versions were given at the conference on "Developmental Systems, Competition and Cooperation in Sociobiology and Economics", sponsored by the Stiftung Forschungsinstitut für Philosophie Hannover, in Marienrode, Germany, April 26-28, 1996, and at a conference on Evolution and Human Behavior, sponsored by the Philosophy Department of Stanford University, May 24, 1996. Conversations at both gatherings were both stimulating and useful.

References

BATESON, P.: "Review of *The Selfish Gene"*, *Animal Behaviour*, 26 (1978), pp. 316-318.

BATESON, P. P. G.: "Sociobiology and Human Politics", in: S. ROSE, L. APPIGNANESI (Eds.): *Science and Beyond*, Oxford (Blackwell) 1986, pp. 79-99.

BATESON, P.: "The Biological Evolution of Cooperation and Trust", in: D. GAMBETTA (Ed.): *Trust: Making and Breaking Cooperative Relations,* Oxford and Cambridge, MA (Blackwell) 1988, pp. 14-30.

BATESON, P.: "Are There Principles of Behavioural Development?, in: P. BATESON (Ed.): *The Development and Integration of Behaviour.* Cambridge (Cambridge University Press) 1991, pp. 19-39.

BRANDON, R. N., ANTONOVICS, J.: "The Coevolution of Organism and Environment", in: G. WOLTERS, J. G. LENNOW (in collaboration with P. MCLAUGHLIN) (Eds.): *Concepts, Theories, and Rationality in the Biological Sciences*, Pittsburgh, PA (University of Pittsburgh Press) 1995, pp. 211-232.

CALLON, M.: "Techno-economic Networks and Irreversibility", in: J. LAW (Ed.): *A Sociology of Monsters: Essays on Power, Technology and Domination*, Sociological Review Monograph 38, London (Routledge/Kegan Paul) 1991, pp. 1322-161.

CAPORAEL, L. R., DAWES, R. M., ORBELL, J. M., VAN DE KRAGT, A. J. C.: "Selfishness Examined: Cooperation in the Absence of Egoistic Incentives", *Behavioral and Brain Sciences*, 12 (1989), pp. 683-739.

DAWKINS, R.: *The Selfish Gene,* Oxford (Oxford University Press) 1976.

DAWKINS, R.: *The Extended Phenotype*, Oxford (Oxford University Press) 1983.

DENNETT, D. C.: *Darwin's Dangerous Idea*, New York (Simon/Schuster) 1995.

FUJIMURA, J.: "Crafting Science: Standardized Packages, Boundary Objects, and "Translation", in: A. PICKERING (Ed.): *Science as Practice and Culture*, Chicago and London (University of Chicago Press) 1992, pp. 168-211.

GODFREY-SMITH, P.: *Complexity and the Function of Mind in Nature*. Cambridge: (Cambridge University Press) 1996.

GOFFMAN, E.: *The Presentation of Self in Everyday Life*, New York (Doubleday) 1959.

GORDON, D.: "The Evolution of Social Behavior", in: R. SINGH, C. KRIMBAS (Eds.): *Evolutionary Genetics from Molecules to Morphology*, Cambridge (Cambridge University Press) in press.

GOTTLIEB, G.: "Development of Species Identification in Ducklings: IV. Change in Species-specific Perception Caused by Auditory Deprivation", *Journal of Comparative and Physiological Psychology,* 92 (1978), pp. 375-387.

GRAY, R. D.: "Metaphors and Methods: Behavioural Ecology, Panbiogeography and the Evolving Synthesis", in: M.-W. HO, S. W. FOX (Eds.): *Evolutionary Processes and Metaphors,* London (Wiley) 1988, pp. 209-242.

GRAY, R. D.: "Death of the Gene: Developmental Systems Strike Back", in: P. E. GRIFFITHS (Ed.): *Trees of Life: Essays in Philosophy of Biology,* Boston (Kluwer) 1992, pp. 165-209.

GRIESEMER, J.: "The Informational Gene and the Substantial Body: On the Generalization of Evolutionary Theory By Abstraction", in: N. CARTWRIGHT/M. JONES (Eds.): *Varieties of Idealisation*, Amsterdam (Poznan Studies in the Philosophy of the Sciences and the Humanities, Rodopi) in press.

GRIFFITHS, P. E., GRAY, R. D.: "Developmental Systems and Evolutionary Explanation", *The Journal of Philosophy*, 91 (1994), pp. 277-304.

GRIFFITHS, P. E., GRAY, R. D.: "Replicator II: Judgement Day", in prep.

HINDE, R. A.: "Dichotomies in the Study of Development", in: J. M. THODAY, A. S. PARKES (Eds.): *Genetic and Environmental Influences on Behaviour,* New York (Plenum) 1968, pp. 3-14.

HO, M.-W.: "On Not Holding Nature Still: Evolution by Process, Not by Consequence", in: M.-W. HO, S. W. FOX (Eds.): *Evolutionary Processes and Metaphors,* Chichester/New York (John Wiley/Sons) 1988, pp. 117-144.

HULL, D.: *Science as a Process*, Chicago (Chicago University Press) 1988.

INGOLD, T.:"'People Like Us': The Concept of the Anatomically Modern Human", *Cultural Dynamics*, 7 (1995), pp. 187-214.

JOHNSTON, T. D.: "The Persistence of Dichotomies in the Study of Behavioral Development", *Developmental Review,* 7 (1987), pp. 149-182.

JOHNSTON, T. D.: "The Influence of Weismann's Germ-Plasm Theory on the Distinction between Learned and Innate Behavior", *Journal of the History of the Behavioral Sciences*, 31 (1995), pp. 115-128.

JOHNSTON, T. D., GOTTLIEB, G.: "Neophenogenesis: A Developmental Theory of Phenotypic Evolution", *Journal of Theoretical Biology*, 147 (1990), pp. 471-495.

KELLER, E. F.: *Secrets of Life, Secrets of Death,* New York (Routledge) 1992.

KELLER, E. F., LLOYD, E. A. (Eds.): *Keywords in Evolutionary Biology*, Cambridge, MA (Harvard University Press) 1992.

KITCHER, P.: *Vaulting Ambition; Sociobiology and the Quest for Human Nature,* Cambridge, MA. (MIT Press) 1985.

KOHN, A.: *The Brighter Side of Human Nature*, New York (Basic Books) 1990.

LATOUR, B.: *We Have Never Been Modern.* (Catherine Porter, Trans.) Cambridge, MA (Harvard University Press) 1993.

LEHRMAN, D. S.: "Semantic and Conceptual Issues in the Nature-Nurture Problem", in: L. R. ARONSON, E. TOBACH, D. S. LEHRMAN, J. S. ROSENBLATT (Eds.): *Development and Evolution of Behavior,* San Francisco (Freeman) 1970, pp. 17-52.

LEWONTIN, R. C.: "Gene, Organism and Environment", in: D. S. BENDALL (Ed.): *Evolution from Molecules to Men,* Cambridge (Cambridge University Press) 1983, pp. 273-285.

MAYNARD SMITH,J.: "Optimization Theory in Evolution", in E. SOBER (ED.): *Conceptual Issues in Evolutionary Biology* Cambridge, MA (MIT Press/Bradford) 1984, pp. 289-314. Reprinted from *Annual Review of Ecology and Systematics*, 9 (1978), pp. 331-56.

MOSS, L.: "A Kernel of Truth? On the Reality of the Genetic Program", *Philosophy of Science Association Proceedings*, 1 (1992), pp. 335-348.

MOSS, L.: "Genes and Generalizations: Darden's Strategies and the Question of Context", *Biology and Philosophy*, 10 (1995), pp. 483-488.

MOSS, L.: *The Agency of Adaptation,* Ph.D. dissertation, Northwestern University, in prep.

NEUMANN-HELD, E.: "The Gene is Dead: Long Live the Gene", in this volume, pp. 105-137.

NIJHOUT, H. F.: "Metaphors and the Role of Genes in Development," *BioEssays* 12 (1990), pp. 441-46.

NITECKI, M. H. (Ed): *Coevolution*, Chicago (University of Chicago Press) 1983.

OYAMA, S.: "A Reformulation of the Concept of Maturation", in: P. P. G. BATESON, P. H. KLOPFER (Eds.): *Perspectives in Ethology,* Vol. 5. New York (Plenum), pp. 101-131.

OYAMA, S.: *The Ontogeny of Information,* Cambridge (Cambridge University Press) 1985.

OYAMA, S.: "Stasis, Development and Heredity", in: M.-W. HO, S. W. FOX (Eds.): *Evolutionary Processes and Metaphors,* London (Wiley) 1988, pp. 255-274.

OYAMA, S.: "Innate Selfishness, Innate Sociality", *Behavioral and Brain Sciences,* 12 (1989) pp. 717-718.

OYAMA, S. (1991a): "Bodies and Minds: Dualism in Evolutionary Theory", *Journal of Social Issues,* 47 (3) (1991), pp. 27-42.

OYAMA, S. (1991b): "Essentialism, Women and War: Protesting Too Much, Protesting Too Little", in: A. E. HUNTER (Ed.): *Genes & Gender VI, On Peace, War, and Gender,* New York: (Feminist Press) 1991, pp. 64-76. Reprinted in M. GERGEN, S. DAVIS (Eds.): *Toward a New Psychology of Gender: A Reader,* London (Routledge) in press.

OYAMA, S: "The Accidental Chordate: Contingency in Developmental Systems", *South Atlantic Quarterly,* 94 (1995), pp. 509-526.

OYAMA, S.: "Locating Development, Locating Developmental Systems", in: S. GELMAN, P. MILLER, K. NELSON, E. K. SCHOLNICK (Eds.): *Conceptual Development: A Piagetian Legacy,* Hillsdale, NJ (Erlbaum) in press.

RAFF, R. A., KAUFMAN, T. C.: *Embryos, Genes, and Evolution: The Developmental-Genetic Basis of Evolutionary Change,* New York (Macmillan) 1983.

SAPP, J.: *Beyond the Gene: Cytoplasmic Inheritance and the Struggle for Authority in Genetics,* Oxford (Oxford University Press) 1987.

SHANAHAN, T.: "Kitcher's Compromise", *Studies in History and Philosophy of Science,* in press.

SHAPIN, S.: *A Social History of Truth,* Chicago (University of Chicago Press) 1994.

SMITH, B. H.: *Contingencies of Value,* Cambridge, MA (Harvard University Press) 1988.

SMITH, B. H.: *Belief and Resistance: Dynamics of Contemporary Theoretical Controversy,* Cambridge, MA (Harvard University Press) in press.

SMITH, K. C.: *The Emperor's New Genes: The Role of the Genome in Development and Evolution,* Ph.D. dissertation, Duke University 1994.

SOBER, E.: "Models of Cultural Evolution", in: P. E. GRIFFITHS (Ed.): *Trees of Life: Essays in Philosophy of Biology,* Boston (Kluwer) 1992, pp. 17-39.

STERELNY, K., SMITH, K. C., DICKISON, M.: "The Extended Replicator", *Biology and Philosophy,* 11 (1996), pp. 377-403.

TAYLOR, P.: "Community", in: E. F. KELLER, E. A. LLOYD (Eds.): *Keywords in Evolutionary Biology,* Cambridge, MA (Harvard University Press) 1992, pp. 52-60.

TAYLOR, P. J.: "Building on Construction: An Exploration of Heterogeneous Constructionism, Using an Analogy from Psychology and a Sketch from Socio-Economic Modeling", *Perspectives On Science,* 3 (1995), pp. 66-98.

VAN DER WEELE, C.: *Images of Development: Environmental Causes in Ontogeny,* Ph.D. thesis, Free University of Amsterdam, 1995.

WEBER, B. H., DEPEW, D. J.: "Does the Second Law of Thermodynamics Refute the Neo-Darwinian Synthesis?" in this volume, pp. 50-75.

WEISMANN, A.: *The Germ Plasm: A Theory of Heredity*, New York (Scribner's) 1893.

WILLIAMS, G. C.: *Adaptation and Natural Selection*, Princeton (Princeton University Press) 1974.

WIMSATT, W.: "Reductionistic Research Strategies and their Biases in the Units of Selection Controversy", in: E. SOBER (Ed.): *Conceptual Issues in Evolutionary Biology,* Cambridge, MA (MIT Press/Bradford) 1984, pp. 142-183. Reprinted from T. NICKLES (Ed.): *Scientific Discovery, Vol II: Historical and Scientific Case Studies*, Dordrecht (D. Reidel) 1980, pp. 213-259.

WIMSATT, W. C.: "Developmental Constraints, Generative Entrenchment, and the Innate-Acquired Distinction", in: W. BECHTEL (Ed.): *Integrating Scientific Disciplines*, Dordrecht (Martinus-Nijhoff) 1986, pp. 185-208.

WOLTERS, G.: "Comment on Brandon and Antonovics' 'Coevolution of Organism and Environment'", in: G. WOLTERS, J. G. LENNOW (in collaboration with P. MCLAUGHLIN) (Eds.): *Concepts, Theories, and Rationality in the Biological Sciences*, Pittsburgh, PA (University of Pittsburgh Press) 1995, pp. 232-240.

Chapter 5

The Gene Is Dead - Long Live the Gene! Conceptualizing Genes the Constructionist Way*

EVA M. NEUMANN-HELD

I. "Genes" and the "Unity of Nature" - Some General Remarks

Mae-Wan Ho (1988) pointed out that it is a particular problem of contemporary biology that it destroys the "Unity of Nature" by looking through the "Neo-Darwinian looking glass". The "Unity of Nature" was, and is, a fact of immediate experience for many civilizations past and present. The history of science indicates, however, that science "is concerned with separating and re-

* For numerous discussions and support I want to thank Prof. Dr. Michael Drieschner, Norbert Held and Romy Tajon. In addition, I am very obliged to Dr. Barbara Alexius for her help in proof-reading the English version of my manuscript.

ducing this unity into ever smaller and smaller fragments of which nature somehow must be glued together." Ho continues: "It is a history, not only of fragmentation, but of our own alienation from nature." Not only the inorganic world has become reduced to mechanisms and atoms, but the organic world as well. In this process it is "Neo-Darwinism which dealt the final blow in disintegrating the organism to a mere collection of particles (genes) shuffled by blind selective 'forces'." The result is that "....humanity [is left] to the dilemma of a disembodied and hence impotent mind pitched against the mindless automation of a body controlled by genes whose sole imperative is to replicate." In opposition to this approach, Ho demands that we return to a rival tradition "that resolutely resists fragmentation in favour of integration and process." Because: "once we begin to see biology again in the light of nature's unity, mind and body will become reunited through processes embracing every level from the sociocultural to the molecular. The organism itself - its functions, volitions and actions - will then be rightly perceived, not as the sole *consequence* of natural selection, but as a focus of being *immanent to process* and emerging *simultaneously* with it." (Ho 1988, p. 117f.). The organism becomes relocated within nature.

Although I sympathize with Ho's point of view, the methodological question has to be asked, does Ho's demand to resist the fragmentation of nature require that "fragments" (e.g. genes) have no place anymore in integrative descriptions of the organism? Are "fragmentation" and "integration" opposites? To approach these questions, one first has to distinguish between an integrative *experience* and an integrative *scientific description*. The latter is what I want to concentrate on, and it is this that Ho as a scientist is working on. Scientific descriptions are given in the framework of concepts and theories, which require a differentiative approach toward the object of description. To describe the properties of an object or a process, which an object runs through, the object has to be "cut into pieces". This is the *fragmentation* or *differentiative* part, and of course, Ho herself distinguishes between environment and organism, and between culture and genes, although "the relationship between organism and environment forms a seamless whole in development" (Ho/Fox 1988, p. 5). A conceptual differentiation is indispensible to make the whole accessible to descriptions as structure or process. The crucial point is, however, that a conceptual fragmentation is not enough. To the contrary, the relationship of the conceptually differentiated entities between each other and to the system has to be worked out. This is the *integrative* part, in which the relations between the "pieces" and their variation in time become part of the description. In the description of organisms (more generally: of systems), biology still has to perform the integrative part. So far, biology can describe organisms down to the molecular level of genes. However, the interactions of genes with other, non-genetic components to form an organism is far from being understood.

Differentiation and integration form two parts of scientific progress. However, they do not just follow each other, but differentiated concepts (and theories) and integrative efforts interact with each other. On one hand, the success of each endeavor of integration is essentially depending on a precise conceptual differentiation preceding the integration. The integrative perspective, on the other hand, can change the underlying concepts. These are exactly the problems which new integrative efforts in biology are facing. These efforts attempt to either complement or even replace the Neo-Darwinian paradigm by a new paradigm, which absorbs results of modern biological research and integrates them in a new holistic (that is: system-related) theoretical framework. One task of this research program of "Developmental System Approach" or "Constructionism"[1] is the critical reflection and the possible new conceptualization of theories. Relevant theoretical concepts have to be identified and specified in the new integrative frame. Such a project requires that we clarify the relation of the system to the entity (structure or process), which is described by the concept. This contextualization allows for reflection of the limits of the discourse within which it makes sense to treat the entity to which the concept is applied as a "separate entity". In this regard, representatives of constructionism have performed important contributions regarding concepts like "fitness", "natural selection", "heredity", "constraints" (cf. Amundson 1994; Gray 1992; Griffiths/Gray 1994; Ho 1988; Oyama 1985).

Regarding a critical, conceptual analysis, however, the gene concept is particularly important, since the gene is a central, for the Neo-Darwinian paradigm probably *the* central concept of biological theories. The historical development of the gene concept is closely linked with the historical development of genetics and of the Neo-Darwinian paradigm. Preceding the foundational phase of genetics in the 20s of our century, the descision was reached to pursue the investigation of genetic processes rather than the investigation of developmental processes (Bowler 1989; Falk 1991, p. 468f.). The exami-

1 I adopt the label "constructionism" from GRAY (1992), p. 175 and note 23, p. 203. He uses it for the new viewpoint on development and evolution, as it was worked out by Oyama and Lewontin.

nation of mechanisms of inheritance was emphasized, whereas questions regarding the ontogenetic development of the inherited material were postponed. From this, a still prevailing understanding developed, according to which processes and structures, which develop ontogenetically, don't have to be explained any further if they can be classified as "inherited" or "inborn". In the traditional Neo-Darwinian paradigm "inherited" means more or less "determined", or at least "predisposed", by genes. In opposition to this understanding the constructionists under the developmental biologists demand the detachment of the "hegemony of the gene" (Falk 1991, p. 468f.). They ask that developmental processes be taken seriously so that it becomes obvious where explanations are required. Gray, for example, proclaimed recently the "Death of the Gene". At the same time, however, he wrote: "In announcing 'The Death of the Gene' I do not wish to argue that genes are not important in development" (Gray 1992, p. 199). The genes are "dead" and yet, they are still "alive"? This obvious contradiction points to a conceptual problem, which even constructionists do not consider thoroughly enough. This problem is based on insufficient differentiations of the gene concepts in biology. Here, the gene is used as a conceptual entity in numerous disciplines, which differ considerably in theoretical and methodological approaches.[2] In particular nowadays, the gene plays a remarkably important role in those biological areas that try to understand processes in the living organism, that is, developmental processes in a wider sense. In this context, embryology and developmental biology have to be mentioned as well as immunology, medicine and, certainly, molecular biology and gene technology. Here the question is, by which means do genes participate causally in the development of phenotypic traits? However, genes also play a central role in evolutionary theory formation. According to Neo-Darwinian understanding it is only the genes which are inherited across generations. A broad fraction of evolutionary theorists claims that genes are the unit of selection. When Gray proclaimed the "Death of the Gene", he disagreed thereby with Sterelny and Kitcher (1988), who pleaded for genic selection in their article "The Return of the Gene". When

2 DALY (1980), p. 81, described it as a "profound theoretical problem", when "the causal developmental gene" and the "population genetical gene" are treated as the same entity. These two gene concepts would be "two incongruent concepts that have converged at the same bit of DNA." He deplores: "...virtually all of us in animal behaviour occasionally lose sight of this distinction, and that is a major reason why nature-nurture disputes are so persistent and perplexing."

Gray does not want to deny the role of genes in development, he is referring to the role of (molecular) genes in the expression of phenotypic traits.

When constructionists criticize the role of the gene in Neo-Darwinian evolutionary biology, they start their argument by criticizing the role of the gene in developmental processes (cf. Goodwin 1994, Griffiths/Gray 1994; Oyama 1985). This strategy of argumentation requires first a conceptualization of the molecular gene concept. Although constructionists have performed imported ground work in this regard, a precise and satisfactory definition of "gene" in constructionist terms is still pending. I want to develop here exactly such a "constructionist gene conceptualization". I will start with the "molecular" gene concept because this is relevant for understanding developmental processes. Following the constructionist reasoning, important insights into the role of the genes in evolutionary processes can then be deduced. However, such an extended dicussion will have to be continued in another place. In the framework of this article I will therefore restrict myself to the analysis of the developmentally relevant gene concept, and in my further discussion it will be only this concept to which I refer. It is my purpose to extend and strengthen the constructionist criticism of the gene concept by working out three points. 1. So far, the constructionist criticism on the gene concept is not radical enough, but it includes even some "preformationist" attitudes. 2. I will show a "developmental analysis" of processes of polypeptide expression, which will include recent discoveries regarding the molecular organization of the genetic material. The result of this analysis is that there is nothing in the DNA that could be described by the traditional gene concept. (The Gene is Dead). 3. I will discuss recent suggestions for a new gene concept and compare them with my own suggestion of a new (constructionist) gene concept. Furthermore, I will show how my concept can remove some interpretation problems in contemporary molecular biology and medicine. (Long live the Gene).

II. The Constructionist Criticism of the Classical-Molecular Gene Concept

In the conceptual framework of classical genetics a methodological approach developed that allowed for characterization of units of inheritance, the so-called genes. The methods of classical genetics do not allow for an immediate access onto, and therefore no analysis of, the genetic material itself. Nevertheless, the material[3] gene concept characterized the view under which the analysis of molecular structures and processes was started. Based on a series of biochemical and physical investigations, Watson and Crick (1953a, 1953b) developed a model for the structure of deoxyribonucleic acid (DNA). Considering the postulated structure of DNA, they speculated that DNA should be able to fulfill all structural requirements that would have to be expected from the genetic material. The models, which had been worked out in classical genetics, became seemingly confirmed by the research of the following decades. It seemed possible to define and experimentally identify a gene as that segment on the DNA which is coding for a linear polypeptide chain. To each function (polypeptide) there is a discrete material structure on the DNA. I call this definition the classical-molecular gene concept.

From this conceptualization up to this day, the wide-spread, paradigmatic conviction of classical genetics rules in biology that the DNA contains informations in form of genes which exist as some kind of blueprint or program before any kind of polypeptide expression[4] and even before or independently of the system of the organism (or of the cell or the tissue). In this conceptual framework, it is the program (the genes) alone that is transmitted transgenerationally and makes sure that "man beget only man"[5]. Consequently, the program (the genes) determine the phenotype, although, besides the genetic source of information, the influence of environmental[6] conditions is not denied. However, its causal effect on the development is ranked secondary to the genetic program. The description of the causal influence of genes on the phenotype is mainly following what McCain calls the "hedge-

3 Numerous books and articles have been published regarding the history of the gene concept. Here I want to mention only one article which investigates the ontological aspect of the gene concept (between instrumental and material) in its historical development and in its contemporary application: FALK (1984).

4 The term "(polypeptide) expression" is used here to mean the entire process that leads to the synthesis of a polypeptide chain, starting from a DNA-sequence.

5 This notion goes back to ARISTOTLE: "ἄνθρωπος γὰρ ἄνθρωπον γεννᾷ" Met. Z, $1032^{a}25$.

6 In this article, I use "environmental" and "developmental" conditions in the same sense, referring to non-genetic conditions, which influence phenotypic expression.

less hedge" tactic. He describes: "In the 'hedgeless hedge', one admits the existence of an anomaly or problem of theory and then proceeds as though one had not. If one is then accused of neglecting the anomaly, one then produces the admission of its existence as conclusive evidence of one's innocence of the charge." (McCain 1980, p. 126)

This "tactic" can become very problematic for the analysis and interpretation of the development of phenotypes and their possible variations. One example is the standard model of mainstream biology for the development of behavioral traits. According to this model, Gray (1992) summarizes, environmental effects produce learned behavior, whereas genetic information is responsible for innate behavior. As a consequence of this model the impression comes up, so Gray, that any behavior that is present at birth does not require any experiential input for its appropriate development, and furthermore, that such behavior could not undergo modifications by succeeding experiences. This impression leads to the wrong and fatal belief that developmental processes can be - more or less - neglected. Constructionists criticize this dichotomous model not only on the level of behavioral development, but in regard to any phenotypic trait, even including the ontogenetic development of morphological structures (Oyama 1985, p. 127). Furthermore, for constructionists all attempts to characterize phenotypic traits as *natural, genetic, innate, genetic predisposition etc.* are out-of-date interpretations in the sense of "preformationist" theories, which imply a causally effective entity beyond the system of the organism. The *Homunculus* in the brain, who is supposed to be responsible for mental activity, is opposed or even substituted by a *Homunculus* in the so-called genetic program (or better: *as* so-called genetic program)[7, 8].

The complexity of developmental processes, however, argues against such simplifying classification. Coming back to the behavioral level, experiments have shown that so-called "innate" or "genetically determined" behavior does not develop autonomously at all, but rather, requires experiental inputs. Gray ascertains: "These inputs are not just secondary and supportive but *'positive,*

7 OYAMA (1985) has analyzed in detail the metaphors in scientific language, which cement the nature/nurture dichotomy. She records that these metaphors instead conceal the requirement for explanation of ontogenetic (and therefore finally phylogenetic and evolutionary) processes.

8 REHMANN-SUTTER (1993/1994), p. 143, calls these preformationist interpretations "Vorbildtheorie der DNA" ["Model-Theory of DNA", translation by the author].

informative and *constructive'* "(Gray 1992, p. 170). Therefore, these "inputs" are not secondary conditions, but they are causally as effective and influential as genetic conditions. These insights call for an alternative description of developmental processes which are looked for in the framework of constructionism. Constructionists do not claim that phenotypic traits are only learned and triggered by environmental stimuli, but they want to demonstrate that the dichotomous categorization is misleading. Gray argues: "...upon developmental analysis the two categories 'innate' and 'learnt' behaviour expand, interramify and coalesce, and thus cease to be meaningful. *All* phenotypes are the *joint* product of internal and external factors. Development is one process requiring many inputs linked together by complex, non-linear, dynamic systems. It cannot be realistically chopped into two." (Gray 1992, p. 171)[9]. Naturally, genes play an important role in developmental processes. However, they have to be viewed as one kind of developmental resource, their role being not more unique than the role of many other factors. Griffiths and Gray state: "The genes are just one resource that is available to the developmental process. There is a fundamental symmetry between the role of the genes and that of the maternal cytoplasma, or of childhood exposure to language. The full range of developmental resources represents a complex system that is replicated in development. There is much to be said about the different roles of particular resources. But there is nothing that divides the resources into two fundamental kinds. The role of the genes is no more unique than the role of many other factors." (Griffiths/Gray 1994, p. 277).

This description leads to the supposition that genes are just available to the organism like the sun or nutritional resources are, or maybe even more so. In six points, Gray (Gray 1992, p. 175-179) specifies the role of the

9 Clearly, these concepts, for which "inheritance" is one example, require clarification (cf. OYAMA, this volume). I believe it is important to focus on the particular questions asked. One should abstain from distinguishing between the categories "innate" and "learnt" or "caused by environmental conditions", when trying to understand the *development* of any trait. However, in another context it might make sense to distinguish between these categories, for example, when asking for the cause of a *difference* between two traits. This difference *could* be caused by a particular part of the DNA, as it was passed down from parents to offspring (inheritance in a narrow sense) (Neumann-Held 1997). However, for the development of the trait the DNA is not sufficient, and, furthermore, under different developmental conditions, the trait could develop differently, as well.

genes, as it is conceptualized so far in the constructionist framework: 1. Genes and the developmental context jointly determine the phenotype. 2. The effects of genetic and environmental variations depend on the context, in which they are happening. Therefore, they are not only reciprocally contingent, but also temporally dependent. The effect of an environmental factor varies with the developmental state of the organism, just as the effect of a gene varies with the state of the entire system. 3. Internal (genes) and external (environmental conditions, non-genetic influences) factors are not independent variables, and they do not exist independently from each other. To the contrary, they *co-define* and *co-construct*. The environment of an organism plays an essential role in the construction of a phenotype, and the organism selects and modifies (constructs) this environment according to the available resources. A gene "does" something only, if it is imbedded in an appropriate developmental context. "Thus", Gray summarizes: " ....a gene can only be functionally defined in a specific developmental context." (Gray 1992, p. 176). 4. All phenotypes are constructed, not transmitted. Neither blueprints nor programs are inherited, but all phenotypes have to develop through organism - environment transactions. Exactly this explains the title of Susan Oyama's book "The Ontogeny of Information" (Oyama 1985). Information is neither *in* the genes nor *in* the environment, but develops (ontogenetically) in the fluid, contingent *relation* between both. 5. Since information is not localized in the genes the control of developmental processes cannot be in the genome. Instead it is dissipated or disseminated over the entire developmental system. Therefore, genes are no "hegemonial" controllers of development, but rather one interacting component in a field of context-dependent difference makers. The information in such a system is always in-circuit and not localized in internal or external factors. 6. Consequently, the concept of inheritance has to be considered again. Not only genes are inherited, but the entire developmental system, including numerous non-genetic entities.

This critique of the description of the role of genes in developmental processes is mainly directed toward the functional aspect of the classical-molecular gene concept. A gene can be functionally defined only in a specific developmental context, a gene as such does not contain any information, genes are not autonomous, but even the replication of DNA requires basically the entire system of a cell.[10] There are good reasons for this (constructionist) interpretations, and I will provide some evidence later.

The classical-molecular gene concept, however, includes not only a functional, but also a structural aspect (Fogle 1990, p. 357). Even in modern textbooks of molecular biology a gene is defined as a certain segment (that might be interrupted) of the DNA, which has the function to code for a linear polypeptide chain, regardless of how complex the mechanism of expression might be.[11] Regarding its functional aspect, constructionists accuse the classical-molecular gene concept of preformationist tendencies. They emphasize that a functional definition of the gene must be context dependent. Constructionists overlook, however, the preformationist tendencies in the *structural* sense. Just this segment on the DNA is further treated as a structured entity, as a gene, which in a certain sense seems to exist independently of the system of the organism (or even before the system), and which is only activated by the system.

With this position, the constructionist criticism of the gene concept runs into problems. It is quite possible to argue that, if a gene is a unity in a structural sense, then there is in the structural unit implicitly a functional unit as well.[12] Although Gray, for example, argues that "there is no simple correspondence between the sequence of DNA base pairs and the functional ac-

10 This last point of citicism was stressed not only by GRIFFITHS/GRAY (1994), p. 299, but also by others, for ex.: LEWONTIN (1993), p. 48, and REHMANN-SUTTER (1993/1994), p. 146.

11 For textbook examples, see: LEVIN (1985), p. 685, and KNIPPERS (1997), p. 28, p. 283f.

12 I assume that this is the reason why REHMANN-SUTTER (1993/1994) locates - in some sense - the information for the polypeptide expression in the DNA. He writes: "Die Gene ....sind ein organisiertes Lager von codierten Proteinsequenzen....". (p. 148). ["The genes ...are an organized store of protein sequences..."]. Although Rehmann-Sutter stresses that the processes of polypeptide expression depend on the "....Gesamtzustand der Zelle und ihrer Umgebung...." ["...state of the entire system and its environment..."] (p. 148), he does not oppose a definition of the gene as structural entity. Rather, he writes: "Die Zellen sind so eingerichtet, daß ihre ständige innere Interaktion mit Genen die Prozesse *möglich macht*, welche zum Leben eines Lebewesens gehören." (p. 147f) ["The cells are made in such a way that their permanent inner interactions with genes make possible the processes which belong to the life of a living being"] [Translations by the author].

tivity of the proteins they are claimed to 'code' for"[13], he grants on the other hand that "the nucleotide sequence does specify the primary structure of a protein" (Gray 1992, p. 170). Obviously, Gray still assumes a (more or less) simple correspondence between a DNA segment and a linear polypeptide chain. Similarly, Oyama states: "...it makes sense in general to say that the primary structure of a polypeptide is encoded on the chromosomes...." (Oyama 1985, p. 70). The question is, of course, in which sense that makes sense. There seems to be an agreement that the co-defining and co-constructing processes of phenotypic trait development first start above the structural level of genes. Using a metaphor, Gray illustrates the constructionist view on the functional aspect of the gene concept. He compares the DNA with a literary text, which is "composed by rows of letters linked into smaller and larger units..." (Gray 1992, p. 177) (the structural aspect of the gene). The meaning of these units, however, is not intrinsic to the units, but "is constructed in the context of a reader of the appropriate culture and experience..." (functional aspect). It is exactly in this metaphor that the constructionist concession to the structural aspect of the gene concept becomes obvious.

I want to show, however, that the constructionist critique of the gene concept can be made much stronger by considering in which ways genes "exist", if at all. According to my following analysis, it is not necessary to make any concessions to the structural aspect of the gene concept, given the most complex expression processes in eukaryotes and most viral systems.

III. The Molecular Mechanisms of Polypeptide Expression

In general, expression processes in a system are described in such a way that all information needed for the expression of a linear polypeptide chain

13 A functional polypeptide has to be folded into a specific three-dimensional structure, and it might have to combine with other polypeptides or co-factors. The information needed for these procedures is certainly not located in the DNA. However, I am interested here in a lower level in the hierarchy of polypeptide expression. I want to discuss the way in which the primary, linear polypeptide sequence is structurally determined by a nucleotide sequence.

seems to be located in the DNA. This "system" can be an organism, a cell or an in-vitro system. Genes are described as DNA segments coding (alone) for linear polypeptide sequences, whereas environmental influences are seen as triggers. In contrast to such a view, even the description of less complicated processes of transcription and translation calls for describing the DNA as only one constituent of a system, which, in addition, includes numerous non-genetic components. I will present such a comprehensive view of polypeptide expression, whereby it will become clear that the non-genetic components participate causally in processes of polypeptide expression.

1. The Process of Transcription

Usually, the description of the processes by which a polypeptide is expressed starts with the DNA, and continues with describing the processes by which one single, complementary strand of mRNA is synthesized from one of the two strands of DNA.[14] This synthesis depends on numerous factors. Specific enzymes of the cell recognize different locations on the DNA as signals relevant for transcription. The mRNA polymerase, for example, which is catalyzing the synthesis of the complementary mRNA, binds to the so-called *promoter,* a DNA site proximal to the transcription initiation site. Often, the promoter has the characteristic base sequence: TATA. Furthermore, there are more nucleotide sequences, which influence transcription of the DNA. These are, for example, the so-called CAAT-region, the GC-box, which, contrary to TATA-boxes, can promote transcription in both directions, so-called *enhancers,* segments of nucleic acids, which can vary the transcription rate from a distance of several thousand base pairs, and sequences that indicate the end of the transcriptional process to the mRNA polymerase. A description of structural and functional elements on the DNA in such a way runs the risk of viewing those elements as being parts of the DNA independently of the system. However, these segments on the DNA only become "carriers of meaning" in interaction with the system of the cell and the organism, respectively. The "significance" that is assigned to DNA segments through transcription processes depends on the system, as becomes obvious from the following

14 There are some biochemical differences between RNA and DNA. For example, thymidine in DNA is replaced by uracile in RNAs. However, for my analysis, this and other differences can be neglected.

points. 1. Numerous polypeptides are not expressed continuously, but rather their orderly expression, and therefore their orderly employment in the system, makes them important metabolic tools. This regulation[15] depends on environmental factors, to which daylight and nutritional resources belong as well as the system's developmental (in its life cycle) and metabolic state. In these cases, so it can be concluded, only distinct DNA sequences are used temporaryly as regulatory entities in the cell. 2. The composition of nucleic acid sequences, which characterize regulatory DNA-segments, are not limited to those segments. T, A und G, C rich regions are not only present in the TATA-box and the GC-box, respectively. They can be found in other DNA regions as well, which do not exhibit regulatory functions. On the other hand, a region can become recognized as, for example, promoter at one time, whereas the same region in the same organism exhibits a different function at another time, for example, as part of a "coding region"[16], as is the case with so-called "overlapping genes". It follows that the developmental context determines the function of nucleic acid sequences in the cell. 3. This context dependency has more consequences. Whether or not a distinct base sequence at a distinct time in a distinct state of the system under distinct environmental conditions is recognized by the transcriptional system as a promoter cannot be deduced from the existence of a distinct nucleic acid sequence. It is only reliable to call a distinct base sequence a promoter when it can be shown in an experiment that this sequence works as a promoter. The same applies to all regulatory sequences on the DNA. 4. Furthermore, there are numerous cases in which, in addition to the promoter, so-called *cis*-acting elements can be identified proximal to the transcription initiation site. All these elements can influence the transcription rate. However, these elements do not act as discrete, independent units. To the contrary, these elements can have qualitative and quantitative effects that depend in non-additive ways on the presence and the effects of other *cis*-acting elements.[17] 5. The same applies to the genomic environment, in which the transcribed DNA-segment is located. It was already shown in the 30s that variations in a certain phenotypic trait can be due to

---

15 The transcriptional as well as the translational rate are regulated this way.

16 "Coding region" is the region whose nucleic acid sequence becomes the template for the synthesis of the polypeptide sequence during the processes of expression.

17 The transcription rate can also be changed by DNA segments distal to the transcribed region. For examples, see: FOGLE (1990), p. 359f., and in particular: LEE et al. (1987).

different relative genomic positions of a certain DNA-segment, from which expression of the phenotypic trait had started (the so-called "position-effect")[18].

All five points, and this list does not claim completeness, indicate that the functional aspect of so-called regulatory sequences is reciprocally contingent on the developmental context, as Gray ascertained more generally for the entire gene. Of particular interest for my argument are, however, points one to three. They allow the argument that the *structural* aspect is context dependent as well! To follow this argument, what "structural aspect" refers to must be understood. Certainly this term means more than only the sequence of nucleic acids on the string of DNA. If one talks about a structure on the DNA (like a promoter), one makes a demarcation by which, in the *description,* this particular segment of DNA is separated from other segments of the linear nucleic acid sequence. This segment has a beginning and an end, and inside those boundaries one finds a more or less characteristic composition. The sequence GTCCATATAAGA as such is an unstructured nucleic acid sequence. Only when the reader performs some structuring and "sets" or "describes" a beginning at the sixth nucleic acid base, a putative promoter region comes into sight. The structure:....GTCCA-TATAA-GA.... of the promoter *comes into being* upon knowledgeable reading. "Reading" in the organism means that the system uses this sequence as promoter in a developmentally specified context. In a different context the system might read the structure: ...GTC-CAT-ATA-AGA..., and translate this structure into the amino acids:...Valin - Histidin - Isoleucin - Arginin..., according to the so-called genetic code (see: chapter III.2).

At this point, a first result can be summarized. The process of transcription binds the DNA in a larger system context, in which parts of the system interpret parts of the DNA as a *structure,* in a way, which is contingent upon the developmental context. The structure then has a certain function (as a regulatory element) in the process of transcription. These structures are not defined before the system, but they come into being in the process of transcription. The developmental context includes the physiological and metabolic state of the organism as well as geographical conditions, social space, and

18 For examples, see: FOGLE (1990), p. 358f., and in particular: LEVIS et. al. (1985).

cultural traditions[19]. The particular states depend on and are limited by the preceding interactions of organism and environment, which have determined the present state. The result is a transcription rate, regulated in a distinct way. In constant ways DNA, enzymes and other environmental factors, which causally participate in the regulation of these processes, interact with each other. What creates this constancy in *these* organisms of *this* kind in *this* situation? The explanation can obviously not be looked for in the "program" of the DNA, as it has been done traditionally, because independently of context and system, the DNA has neither structure, nor function, nor program, nor information. Rather, the constancy of the transcriptional processes has to be attributed to the constancy of the patterns of interaction of the participating components. Constructionists have pointed to the constancy of patterns of interacting components[20] to explain numerous developmental processes so far. My discussion indicates, however, that the constancy of patterns of interacting components also explains the process of "structurization" of regulatory sequences in the process of transcription. No basic program is required, but the process is self-organizing on the basis of the preceding steps and of the components (be it DNA-sequences or sun light) that are available to the system. If one of the components changes, the result of the process can change as well. This explains, for example, why a change (mutation) in a promoter region can leave the promoter unrecognizable or can change the transcription rate. The same effect can be caused by a change in the environment, be it the genomic environment or the - regarding the organism - outer environment. Of course, these considerations do not apply to promoters only. The basic DNA sequence and the developmental context determine in reciprocal contingency the structure (and function) of all regulatory sequences of transcription or translation; they *co-define* und *co-construct.*

Does this, however, also apply to "coding regions"? Are not the "coding regions" the real carriers of information, according to the classical-molecular concepts? (cf. Carlson 1991, p. 478). And when constructionists ascertain that genes are functionally defined only in a context, they certainly refer (also) to the coding regions and not (only) to regulatory regions. Therefore, the ac-

---

19 In plants, even touch can belong to these environmental contexts (BOYD [1990], p. 40).

20 *Interactive constancy* is in addition to *co-dependency, reciprocal contingency* and *co-construction* a further term that was developed in the constructionist conceptual framework for the explanation of developmental processes. Cf. GRAY (1992), p. 181f.

tual touchstone for my argument is the question: Are "coding regions" also structurally defined (and developmentally constructed) in a way that is contingent upon the developmental context? This is exactly the point for which I want to argue, based on the observation of so-called post-transcriptional processes. Therefore, we must focus on those processes which follow the completed process of transcription. For the upcoming discussion, however, understanding should be facilitated by some short remarks regarding the processes of translation.

2. The Process of Translation

The term "translation" indicates the process by which linear polypeptide chains are synthesized on the basis of a mRNA. The translating system assigns to every three succeeding nucleic acid bases (called: *triplets* or *codons*) of the mRNA one particular amino acid. Numerous molecules participate in this process. In particular, these are ribosomes, complex molecules that catalyze the translation of mRNA sequences into amino acids, and classes of tRNAs. *t* stands for *transfer* and indicates that these molecules transfer specific single amino acids to specific mRNA sequences, where the amino acids are joined to form a polypeptide chain. The transfer of the order of nucleic acids in the mRNA into the order of amino acids in the synthesized polypeptide is guaranteed by the fact that the tRNAs bind by a specific triplet combination to the complementary sequence of the mRNA. Each class of tRNAs possesses one particular triplet combination at a certain site in the molecule, which corresponds with the ability to bind to a particular amino acid at another site. The triplet AUG on the mRNA can only bind to a tRNA, which possesses the complementary triplet UAC. This tRNA is at the same time a carrier of the amino acid methionine. By binding of the corresponding tRNAs to the complementary triplets of the mRNA, the corresponding amino acids get into spatial proximity, so that enzymes can catalyze a chemical linkage between them. Thereby, the binding to the carrying tRNA is resolved enzymatically. The first links of the polypeptide are established.

This is, of course, only a very short outline of the extremely complicated processes of translation. However, the reciprocal contingency of the participating components becomes quite obvious. The description of the significance of the triplets indicates clearly that it is essential for the process of translation that the ribosomes find the "correct starting-point". The first nucleic acid of the first triplet must be recognized as well as the correct end. If the translation process starts or ends at alternative sites on the mRNA, or if there is a shift in the reading frame for other reasons (for example: an insertion or deletion of nucleic acids), then this will lead to a different polypeptide chain or to an early translation stop. It is all the more astonishing that in living systems such alterations happen quite frequently in posttranscriptional processes on the mRNA.

3. Posttranscriptional Processes - mRNA Processing

mRNA, the product of transcription, can become submitted to numerous modifications (mRNA processing). In the light of the following examples, the first is probably the most "simple" modification. In eukaryotes, the so-called 5'end of the transcript is "capped" by a methylated Guanine, whereas the other end of the transcript, the 3'end, is shortened by a few nucleotides, whereafter a tail of about 200 adenine nucleic acids is added (polyadenylation). These modifications, which are catalyzed by specific enzymes, are not prescribed in the DNA. However, they are essential for further processing (including translation) of the mRNA in the eukaryotic cell. Therefore, this is another example that DNA is not the sole carrier of information in the process of expression.

More interesting are certain modifications which are frequent in eukaryotes, too. They often have so-called "split genes". This name indicates that after transcription, parts of the original transcript (primary transcript) are removed, the so-called introns. The remaining parts, the exons, are spliced together to form the so-called mature mRNA. Primary transcripts can contain 1 to 50 introns and 90 % of the primary transcript can consist of introns. In many cases, a "reasonable" reading-frame is created only after splicing. Now, one might assume that particular "signals" (prescribed structures in form of specific nucleic acid sequences) on the mRNA indicate to appropriate enzymes where introns have to be cut out. However, this model has to be abandoned in the light of alternative mRNA processing. During alternative mRNA processing different parts of the primary mRNA are cut out. A very impressive example is the gene for the bovine growth hormone. Hampson and Rottman (1987) found that the splicing mechanism distinguishes in one case five exons and four introns (case *a*), whereas in an alternative case only three introns are cut out (case *b*). That segment of the primary mRNA which became the fourth intron in case *a* is part of the coding region in case *b*. By that, the reading frame shifts from that point on, where the fourth intron would have started in case *a*. That segment which is the fifth exon in case *a* is therefore translated in case *b*, too, but into a completely different amino acid sequence. This alternative polypeptide product can be longer or shorter than the translation product of case *a*. Hampson and Rottman conclude that the polypeptide products of case *a* and case *b* have 125 amino acids in common whereas they differ in the remaining 66 to 108 amino acids.

Alternative mRNA processing is surely not a rare phenomenon[21]. It depends on environmental conditions, like, for example, developmental cues of the organism and of the cell type (Fogle 1990, p. 361). Therefore, it is again the developmental system, which, in reciprocal contingency with the primary mRNA, co-constructs in each case the relevant mature mRNA! The consequence of this process of differential intron removal is that differential polypeptide products are synthesized in the following steps. These syntheses, I want to emphasize again, have originated from the same bit of DNA.

At least, one could argue now, the nucleic acid sequence of the DNA is given. As a matter of fact, it is not. Probably the most unbelievable kind of obtaining different polypeptides from the same DNA segment is mRNA editing in mitochondria and chloroplasts. It was described for the first time in 1986 in mitochondrial transcripts of trypanosomes.[22] Meanwhile, mRNA editing was shown in Physarum polycephalum, mammals, viruses and higher plants (Hanson et al. 1996, p. 57). There are two different kinds of mRNA editing. In one kind, nucleotides are inserted into or removed from the mRNA. The second kind converts nucleotides, for example, C(ytosine) in U(racile) (and the other way around) (Hanson et al. 1996, p. 57f.) and adenosine via inosine into guanosine (Herbert 1996, p. 6). mRNA editing is not a rare phenomenon, but in plant mitochondria, for example, every coding region undergoes mRNA editing (Hanson 1996, p. 1). 3 - 15 % of the triplets

21 For more examples, see FOGLE (1990).

22 Not all mechanisms by which different polypeptide products become synthesized from the same DNA segment include mRNA processing. Two examples: mRNAs are always synthesized from one strand of DNA only in a distinct direction (from 3'end to 5'end of the DNA); on this strand of DNA, coding regions can overlap, or mRNAs can become transcribed in opposite directions (using both strands of DNA). In these processes, too, it is the developmental system that "decides" which DNA segments enter into the expression processes.

of mitochondrial mRNA are influenced by editing, and the extent of editing varies within the same species and between species (Hanson et al. 1996, p. 57f). A kind of "position effect" has been observed here, too. Sometimes, a DNA segment can become situated in different locations in the genome by recombination events. It was reported that mRNAs of such DNA segments are differentially edited, depending on the genomic location of the DNA fragment, of which the mRNA was derived (Hanson et al. 1996, p. 60). Since mRNA editing changes codons, amino acids are strung together in the processes of translation differently from what a genomic DNA analysis would have predicted. This situation becomes even more complicated by the fact that mRNA editing does not have to be complete. Most interestingly, though, the proportions of edited and unedited transcripts vary developmentally, as was shown in mammalian systems as well as in trypanosomes (Hanson et al. 1996, p. 62). Again, it is the state of the developmental system that gives the causal clue to the process of mRNA editing. Consequently, differential editing leads again to differential polypeptide products, whose expression processes have started from the same DNA segment. The *rps12* gene, for example, codes for ribosomal proteine S12 and has six (C-to-U) editing sites. Only the product of the completely edited mRNA is used as ribosomal subunit. However, the partly and non-edited mRNAs, respectively, are translated, too, and the authors have good reasons to speculate that these protein products fulfill different functions (Phreaner et al. 1996). This has been shown for the mammalian apoprotein B. Differential mRNA editing leads to two different proteins with different properties in lipid and transport metabolism (Hanson et al. 1996, p. 62). At least theoretically, mRNA editing can play a role in numerous processes and structures in the cell during polypeptide expression. These processes could include: splicing, codon composition, translation initiation or termination, reading frame, mRNA stability or transport and RNA secondary structure (Hanson 1996; Hanson et al. 1996, p. 58; Herbert 1996, p. 8). mRNA editing is probably performed by different mechanisms (Hanson et al. 1996, p. 59). At least in the case of the mRNA for a glutamate neuroreceptor (*GluR-B*) it is even introns that play an "informational role" in editing of the mRNAs (Herbert 1996, p. 7).

Research on mRNA editing has certainly just started and certainly more surprises can be expected. At this point, I would like to emphasize that these processes can only be described in the following general way: environmental (developmental) conditions, primary mRNA, and processes such as mRNA editing, are in reciprocal ways contingent on each other in the determination of the structures that become translated.

mRNA editing is an extreme example that illustrates that it is (conceptually) impossible to identify coding regions as structures which have functions independently of the developmental context and system. Coding regions, too, are co-defined and co-constructed, as was shown for regulatory elements already. There is no reason to assume that these descriptions do not apply in cases where there is neither mRNA editing nor differential mRNA processing nor mRNA processing at all (as in many prokaryotes). In these cases, it is just not as obvious that the DNA and the mRNA, respectively, are not structured in themselves, except for the sequence of nucleic acids, but that the structure is constructed in succeeding processes. Certainly, upon sequencing of a nucleic acid sequence, molecular biologists can *speculate*, where there might be a coding region (or a regulatory region). However, that means that they interpret into the sequence a structure in a similar way as the expressing system in the cell is doing it.

Therefore, an essential result of this analysis can be summarized. Regulatory sequences and coding regions do not exist on the DNA or mRNA independently of the system (or of somebody who can read, which is exhibited by the cell in the "act" of expression). On the contrary, regulatory elements and coding regions are co-constructed (in a structural and functional sense) in reciprocal contingency by the components of the systems in succeeding processes. Therefore, the constructionists' considerations can be complemented. Reciprocal contingent processes of co-definition and co-construction describe the levels below coding regions und elements as well. The result of these processes between DNA and environment (in its widest sense) are coding regions and regulatory sequences.

This result can be described by an up-dated version of Gray's metaphor. Gray compared the DNA with a literary text. My analysis indicates that this text is not structured into smaller and larger units (words and/or sentences), as in Gray's example, but that it is only a sequence of letters, which is structured only by the experienced reader (developmental context) in reciprocal contingency.

IV. What is a "Gene"?

1. The Problem of a Comprehensive Conceptualization of the "Gene"

The analysis of the molecular mechanisms of polypeptide expression shows quite clearly that there is no fundamental way by which the classical-molecular gene concept could be applied to DNA segments. One focuses at the same bit of DNA, and different structures and functions can appear. One focuses on different levels of the expression process (DNA, primary mRNA, mature mRNA, edited mRNA, polypeptide), and again different structures and functions appear. Introns can become exons, which can become promoters, and so on. Regarding the aspect of function, there is no general rule that a particular sequence codes for only one polypeptide. Also, in principle, no discrete material unit segment on the DNA can be identified as coding for (only) one polypeptide — at least not in the sense of the classical-molecular gene concept. Therefore, this gene concept is no longer useful; it is "dead".

It might be surprising, therefore, that molecular biologists refer to genes quite frequently. As a matter of fact, the "gene" is present in almost all scientific and public discussions around molecular biology and *gene* technology. The crucial and confusing point, however, is that biologists mean very different entities when they "talk genes". This point has been stressed recently by a number of authors. Falk, for example, notes: "... a casual glance at the current genetic literature would be enough to reveal that although the term 'gene' is very much in use, it means different things for different people. .....Today the gene is .... a unit, a segment that corresponds to a unit-function, as defined by the individual experimentalist's needs" (Falk 1984, p. 208)[23]. The experimentalist's needs are versatile, indeed. Carlson summarizes: "The gene may be considered at the level of what is transcribed from the DNA or what is translated from the ultimate part of the mRNA that encodes the amino acids in the protein produced by the gene. There will be those who include the introns in the gene, others who will include the reading initiation and accessory DNA promoters and regulatory receptors contiguous to the gene (but not those DNA elements removed from the gene as upstream or downstream enhancers or regulators). When even more complex genes .....are added to this

---

23 See also, for example: BURIAN (1986, p. 37); KITCHER (1992); KITCHER (1982); PORTIN (1993, p. 173).

search for an all-encompassing definition of the gene, the pursuit becomes quixotic. It is important that geneticists recognize the many levels at which genes can be perceived, but it is not helpful to select one of these levels and arbitrarily designate that as the universal definition of a gene." (Carlson 1991, p. 475)

Although it is quite known that it is scientific practice to apply the term "gene" to different structures and processes, there seems to be a general agreement that this procedure poses no problems for the scientific enterprise. Somehow, so the argument goes, scientists deduce from the context of their talk (or their publication), what they mean when they "talk genes". Therefore, it is quite "arguable that the old term gene...is no longer useful, except as a handy and versatile expression, the meaning of which is determined by the context. In that case, the challenge is to devise a new terminology for use when precision is needed." (Portin 1993, p. 208). However, such precision seems to be rarely necessary, as the practice shows. Carlson states, explicitly: "For most...students... and those who are not engaged in research in genetics, the gene in its functional sense is more helpful than the gene in its complex biochemical or molecular sense. One does not need to be an expert in genetic transcription and translation to convey to a family the risks they face when a grandparent is diagnosed as having Huntington's disease." (Carlson 1991, p. 475). However, just such an assessment is the old "hedgeless hedge" argument in disguise. A gene "in its functional sense" is supposed to mean that there is only an informational flow from DNA via mRNA to the protein, which (alone) determines phenotypic traits (Carlson 1991, p. 475). Therefore, whenever someone only wants to say that something is determined "genetically", she or he just refers to the "genes", as is practice these days, anyway. This way, one can abstract from all the complicated mechanisms, which, if they are described comprehensively, need considerations of non-genetic inputs as well. My analysis shows that we have to consider the reciprocal contingency of DNA and environment that together construct coding regions and regulatory regions. Apart from the developmental system there is no informational flow starting from the DNA. Furthermore, if this insight is really understood and taken into considerations thoroughly, this might very well have consequences for the assessment of the risks of so-called inherited diseases or dispositions to inherited diseases.

Therefore, for the sake of scientific precision, I find it necessary to either abstain from "talking genes" in such confusing ways, or to think about a new gene conceptualization. The first question, consequently, is whether the term "gene" is still usefull in the description of molecular processes of polypeptide expression. Two interesting and inspiring suggestions for alternative gene conceptualizations were published recently, and discussing them will prepare my argument for a third suggestion.

The first suggestion, I want to discuss, was made by Waters (1994). Waters finds that the different uses of the term "gene" in molecular biology can be reduced to an underlying concept. This "fundamental concept.....is that of a gene for a linear sequence in a product at some stage of genetic expression" (Waters 1994, p. 178). Waters emphasizes that this concept specifies, *for what* a gene is, namely for a linear sequence in products of genetic expression. An example might clarify what that means. Introns could belong to a gene as well as not, depending on which "particular linear sequence in a product at some stage of genetic expression" scientists are referring to (Waters 1994, p. 179). An intron is part of a gene in view of the process of transcription at the stage of the primary mRNA. It is not part of the gene in the view of the polypeptide chain. It follows that the product of genetic expression which is subject to scientific description must become specified, at least implicitly. And Waters argues that exactly this is the case in the practice of biology since: "Molecular biologists understand what the term 'gene' refers to in concrete situations because the context of discussion implicitly indicates the relevant stage and product of genetic expression." (Waters 1994, p. 179)

By trying, however, to reduce the inflational use of the "gene" in molecular biology to an "underlying concept", Waters creates new problems. In Waters' analysis the impression is forced upon the reader that not only several genes participate in the processes of polypeptide expression, but that these genes come into being on different levels of the expression process. The entire sequence of the DNA, of which a primary mRNA is transcribed, can be called a gene according to Waters (let us call this gene: GENE 1). GENE 2 then, would be the gene which is composed of the spliced exons. Since introns are "recognized" and cut out only at the level of the mRNA, GENE 2 comes into being on the mRNA and would evolve from GENE 1 through processes of transcription and posttranscriptional modifications, respectively.

So far, I have given an account of Waters' position regarding the context dependency of the scientific description. When trying to conceptualize the gene, however, Waters also wants to pay attention (rightfully) to the genomic context (Waters 1994, p. 181), since he wants to include differential mRNA processing in his argument. But then the situation becomes even more complicated, because now, from GENE 1 (situated on the DNA) could evolve several genes on the mature mRNA. Furthermore, Waters' reference to only the *genetic* context is too one-sided. Waters claims that what a gene is would depend on the genetic context. Therefore it would depend on "whether the primary transcript has been appropriately spliced" (Waters 1994, p. 179). Splicing itself, so Waters, depends "on the sequence of nucleotides within the intron and on sequences distant from the gene" (Waters 1994, p. 181). Apart from the fact that Waters does not specify the gene (GENE 1 on the DNA? GENE 2 on the mRNA?), splicing does not depend exclusively on the genetic material, but also on non-genetic influences, as has been discussed.

So far, my criticism of Waters' argument can be summarized into two points: First, although Waters recognizes a certain context contingency, he restricts this contingency either to the scientific description and methodology, or to the genetic context. Thereby, he misses the environmental influences (in the widest sense) that are not coded for by the DNA (or mRNA). Second, efforts to reduce the applications of the term "gene" in molecular biology to an "underlying concept" lead to conceptual problems. The idea is at least very strange that genes on the DNA make genes on the mRNA, and even stranger that a gene on the DNA makes a variety of mRNA genes by mRNA editing.

Waters' suggestion does not bring any conceptual gain in comparison to the ambivalent use of the gene in molecular biology. On the contrary, in molecular biology as well as in Waters' approach, the same facts can be described much more precisely in terms which are already available and used in molecular biology, like non-coding region, coding region, primary mRNA, mature mRNA, intron, exon etc. (although it might be preferable to define these concepts more precisely as well). When describing the processes of polypeptide expression, I have hardly used the term "gene", which proves that this is possible. Interestingly, Waters himself claims that "gene", in the sense he uses it, would be more or less a synonym for "coding region" (Waters 1994, p. 182). But why then use the the term "gene" at all, except for sentimental reasons or out of laziness?

In opposition to Waters, I am convinced that there are good reasons to differentiate conceptually between "coding region" and "gene". Calling a coding region a gene excludes regulatory regions from the gene concept, and Waters explicitly neither counts regulatory regions as self-contained genes nor as part of those genes that they regulate (Waters 1994, p. 180). This deserves some considerations. On one hand, regulatory regions are only regulatory sequences in regard to something they regulate. The object of regulation, however, is mostly the coding region. On the other hand, coding regions, that is, Waters' genes, are only *for* something if they are hooked up with regulatory sequences. Therefore, regulatory sequences and coding regions are intimately connected. The connection of coding region and regulatory sequences is a requirement (not the only one, though) for any kind of polypeptide expression. Therefore, on a certain level of description, it makes sense to coin a term which expresses this necessary connection. What would be an appropriate term for the connected structure of regulatory sequences and coding regions which is conditional for polypeptide expression? I suggest the term "gene" be used, since traditionally, "genes" are thought of as "making" a polypeptide chain anyway.[24] These "genes" are more than coding regions because somehow they bridge the gap between DNA and polypeptides. Of course, the crucial question is, what is meant by "somehow"? Does that mean that genes are somehow glued together of single DNA segments like pieces of a mosaic? It seems that Fogle (1990) thinks in this direction, and therefore, I want to discuss his gene concept as a second suggestion.

In agreement with the view I have presented here, Fogle concludes: "The ..investigation of the unit concept [of classical genetics] finds no feature in common among described protein coding genes of eukaryotes that unites them as material units of structure or function." (Fogle 1990, p. 366). Therefore, it is Fogle's goal: ".... [to] attempt to outline an alternative framework for gene structure mindful that diversity and overlap are normal phenomena requiring integration into any such system." His proposal is "that sets of domains more closely describe a gene, where domain is intended to mean a sequence of nucleotides identified by structural properties and/or activities which distinguish it from other such sequences." (Fogle 1990, p. 367). Promoters, for example, are domains that indicate the transcription start site, and that have a structure in common, which is close to the nucleotide sequence TATA. In Fogle's account, too, the identification of domains is context dependent, so that a nucleotide sequence can be part of more than one domain. This allows him to include "nesting" (a domain is part of a larger domain) and "overlapping". "Exons", Fogle states for example, "have the activity of splicing in common and hence are domains for that process." He continues: "Alternative splicing can change the set of exons or the boundary of some exons while retaining the attendant activity (splicing)." (Fogle 1990, p. 367). With these comments, which bind the characterization of domains to a pro-

24 I admit, however, that it might be worthwhile to consider the use of a different term, since "gene" has such a long history of diverse meanings already.

cess or an activity, Fogle points in a very interesting direction, which, however, he does not think through to an end. I agree with him when he asserts: "A material gene ... lacks coherent meaning at the molecular level until a partial set of transcriptional domains is specified". Therefore, a material gene has to be defined in a relational way.[25] But then, he sticks too much to the level of nucleic acid molecules, and therefore, to the old gene concept, when he states: "A gene looks like a.... collection of component entities that together define its structure and influence the phenotype" (Fogle 1990, p. 367), and furthermore: "... a domain model accepts that information resides within units smaller or larger than a conventionally defined gene..." (Fogle 1990, p. 368). Here, it is not clear what Fogle actually means. It seems that Fogle places the informational content for polypeptide expression back into domains, which he then describes as DNA segments. This is a contradiction to his claim that domains are domains, at least partly, through a process. Fogle runs into these problems because he does not take the contingency of the processes on the developmental context seriously enough. Therefore, he gives at least the impression, domains - as DNA segments - would contain in some "preformationist ways" information independently of the context.

Contrastingly, I have presented arguments that there are no component particles (domains, coding regions, regulatory sequences) on the DNA apart from some developmental context. It is a process that transcriptionally and translationally specifies domains, and this process is constructed in a reciprocally contingent way by the participating components. Therefore, it can be concluded that a gene cannot be a collection of component particles on the DNA as if those particles were "just there". Since these particles or segments are constructed, the gene must somehow be the product of the construction process itself. Certainly, this is quite far away from the traditional view of genetics that would define the gene as a material unit *on* the chromosome. However, as this analysis has shown, the DNA is not structured in a way that allows the fundamental application of the old gene concept. Therefore, the way is clear for thinking about a *new* concept of the gene. This gene concept should capture the idea that "something on the DNA" has to do with the expression of polypeptides, as well as the idea that DNA segments have to be joined together by other non-genetic components, which partly *specify* the DNA segments, required for a specific expression process.

---

25 By the way, WATERS (1994), p. 181, comes to the same conclusion.

2. The "Gene" - An Expanded Constructionist Gene Concept

Usually the view from which expression is investigated starts from the DNA and turns then to the processes which lead to polypeptide(s) as expression product(s). I suggest to reverse this view and look at the *history* of expression of a specific polypeptide. From this point, it is obvious that, by using material, discrete DNA (and mRNA) segments, certain (in principle) experimentally identifiable processes have led finally to this particular expression product at this particular time. This reciprocally contingent interaction between DNA and certain environmental stimuli (a process in itself), which involves interaction of non-genetic components, co-defines certain structures on the DNA. These structures then exhibit certain functions in a thereby specified expression process. During the steps of this expression process the components of the preceding processes and environmental stimuli continue to interact with each other. The result of all these processes is a linear polypeptide chain. I suggest this whole process be called "a gene".

Almost as in the classical sense, a gene then has a function, to code for a polypeptide, and a gene includes certain segments of the DNA. However, in contrary to the classical-molecular concept a *gene is not situated on the DNA*, but the term "gene" describes processes, which under certain environmental conditions structure the DNA (and the mRNA) and result in a polypeptide.[26] The gene becomes relational. Consequently, and this is another essential point, the concept of the gene includes always the interaction between the DNA and the developmental environment. This applies also, as has been shown, to the "subdomains" of the genes. In this context of my description of expression processes it is only the DNA that exists independently of the context.[27] The gene concept, as I suggest it, avoids by definition from taking the DNA as sole informational source, regarding expression of a linear polypeptide. Contrary to Fogle's suggestion, specification of domains is not

26 It should be noted here that not only polypeptides are transcribed from the DNA, but also rRNAs and tRNAs. I do not have a problem calling those processes, which result in rRNAs and tRNAs, respectively, "genes", too.

27 However, the context of description could be changed. The DNA depends on certain environmental conditions, too, like medium of water, salt concentration, pH-value. Furthermore, the DNA sequence can be changed by numerous mechanisms. Transposons und gene-amplification are only some examples. Regarding the fluidity of the genome, see, for example: POLLARD (1988).

sufficient to specify a gene. Rather the concept requires studying also the non-genetic components and processes which participate in the expression and make them possible in the first place.[28] The inclusion and description of these processes and of the relevant environmental conditions - be it metabolic and/or developmental states of the system, be it climatic conditions and/or nutritional resources or even cultural conditions - is certainly extremely complex. However, even this could be understood as an invitation to molecular biology and medicine to reconsider strategies of research and interpretation of empirical results. In the framework of this article, I can only sketch which new approaches develop almost automatically out of this new understanding of the gene. In particular, it is very fruitful to see what happens when the usual use of the term "gene" is substituted by an explicit use of my suggestion. Then, it must be discussed what it means to state that certain genes predispose for certain ailments. In this framework such a statement is meaningless unless it is specified which nucleic acid sequences *and which environmental conditions are referred to,* by talking about a gene for.... More accurately, a medical instruction would have to inform like that: "Owing this nucleic acid sequence, you are endangered, if you put yourself under these environmental conditions..." or "if such a physiological state is reached by consuming those nutritional sources....". This new gene concept urges one to approach the question why an "inherited disease" is not phenotypically present from the very start of life on, but develops at a certain age, that is, contingently on the developmental context. It becomes obvious that a research strategy is unsatisfactory which only identifies "genes" (in the sense of DNA sequences), but neglects (or postpones) research into the mechanisms that make the "genes" effective in connection with non-genetic factors (that is, to ask for genes in the sense specified here). Furthermore, it becomes obvious that a discussion is beside the point which asks whether certain ailments (like allergies) are acquired or inherited. There should be hardly any phenotypic trait

28 Regarding the practice of scientific research, Fogle admits: "Specification of all domains influencing expression may not be necessary for nominative or heuristic purposes and may be impossible to establish empirically." (FOGLE [1992], p. 367) This point applies to my gene conceptualization as well. In addition, not all the environmental conditions have to be (can be) specified. The essential point is that restrictions on *only* genetic or *only* environmental influences have to be given up. It remains a point for a succeeding analysis in another place, to specify the criteria by which necessary influences can be selected.

that is not enabled by genes *in some way*, and which could not be changed by knocking-out some genes. Here again, the essential question is directed toward the interaction of all factors. There can be gene effects that lead to the same phenotype under all known environmental conditions. Traditionally, these cases would be called "innate". But they are the result of constructions in the sense specified here as well, in which the result of expression processes remains stable through different environmental conditions. On the other hand, phenotypic traits can remain quite resistent to genetic changes. Such phenomena are discussed by Brian Goodwin (1994), for example, in detail.

Coming to an end, I want to plead for a restriction in the use of the gene concept specified here. I find it absolutely necessary to restrict the gene concept to those processes that end with the synthesis of a linear polypeptide chain. Starting from here, there are, as indicated, different, mostly non-genetic processes which use in succeeding steps the results of preceding processes as components in further steps. In some sense it is on each new level that new phenotypic traits come into being, which in following levels are integrated with added components into new "higher" phenotypic traits. Linear polypeptide chains become three-dimensional enzymes that can become active in an appropriate cell environment. Under certain circumstances and by adding specific social and/or cultural conditions, these enzymes can even influence human behavior. A mutation in the coding region of the gene *monoamino-oxidase A* prevents, for example, the expression of this gene. Under certain environmental conditions, this can be correlated with amplified impulsive aggressivity in human men (Brunner et al. 1993a and 1993b)[29]. Under different environmental conditions, however, no elevated impulsive aggressivity can be observed. Mutations in the coding region of *monoamino-oxidase A* cannot by itself explain elevated readiness for aggressive behavior, but correspond with it under certain additional conditions. It is a long way from the effect of variations in *monoamino-oxidase A* activity to the phenotype of aggressivity - a way in which more components become influential, such as nutrition, sleeping habits and social environment.

The processes of phenotype development are distributed over numerous levels and are very complex. They are still waiting for conceptualization. This is a further argument not to expand the gene concept to an extent that includes all processes of phenotypic expression. Either that would be a reduc-

29 For more references and a detailed discussion, see: NEUMANN-HELD (1996 and 1997).

tion in its worst sense (everything is only caused by genes), or the gene concept becomes useless. Even today, an extension of what is believed that genes can do is performed much too often, and I find this very problematic.[30]

The goal of this article was to conceptualize genes in a "constructionist way". Thereby, the gene became a product of processes, which are probably products of processes themselves. On the other hand, the gene became component and starting point of numerous processes, in which - level by level - phenotypic traits develop and vary. Probably, these levels do not end with the organism but continue into the description of social systems. From here on numerous questions invite further analysis.

It is our empirical research and our efforts to make sense of the data that force us to think about old concepts and paradigms in biology. I believe, we have to face this challenge, and try to find solutions in order to develop a better understanding of biology, and, as always, of ourselves.

The scientific descriptions of the "Unity of Nature" still has a lot of white spots on the map, and leaves more questions unanswered than have been answered so far. Maybe it is at least occasionally inspiring that apart from all science the "Unity of Nature" can be at least experienced once in a while.

References

AMUNDSON, R.: "Two Concepts of Constraint: Adaptationism and the Challenge from Developmental Biology", *Philosophy of Science,* 61 (1994), pp. 556-578.

ARISTOTELES: *'Metaphysik Z'.* Text, translation and commentary by M. Frede/G. Patzig, München (C.H. Beck) 1988.

BOWLER, P.J.: *The Mendelian Revolution. The Emergence of Hereditary Concepts in Modern Science and Society,* London (The Athlone Press) 1989.

30 Then, the gene for monoamino oxidase A becomes the "gene for aggressivity", or even the "aggressive gene", as happened in publications, cf.: NEUMANN-HELD (1996 and 1997).

BOYD, A.: "Plant Switches on Genes in Response to Touch", *New Scientist*, 28 (1990), p. 40.

BRUNNER, H.G. et al. (1993a): "Abnormal Behavior Associated with a Point Mutation in the Structural Gene for Monoamino Oxidase A", *Science*, 262 (1993a), pp. 578-580.

BRUNNER, H.G. et al. (1993b): "X - linked Borderline Mental Retardation with Prominent Behavioral Disturbance: Phenotype, Genetic Localization, and Evidence for disturbed Monoamino Metabolism", *American Journal of Human Genetics*, 52 (1993b), pp. 1032-1039.

BURIAN, R. M.: "On Conceptual Change in Biology: The Case of the Gene", in: D.J. DEPEW, B.H. WEBER (Eds.): *Evolution at a Crossroads*, Cambridge, Mass./ London (MIT Press) [2]1986, pp. 21-42.

CARLSON, E. A.: "Defining the Gene: An Evolving Concept", *American Journal of Human Genetics*, 49 (1991), pp. 475-487.

DALY, M.: "Contentious Genes: A Commentary on *The Selfish Gene* by Richard Dawkins", *Journal of Social Biological Structure*, 3 (1980), pp. 77-81.

FALK, R.: "The Gene in Search of an Identity", *Human Genetics*, 68 (1984), pp. 195-204.

FALK, R.: "The Dominance of Traits in Genetic Analysis", *Journal of the History of Biology*, 24 (1991), pp. 457-484.

FOGLE, T.: "Are Genes Units of Inheritance?", *Biology and Philosophy*, 5 (1990), pp. 349-371.

GOODWIN, B.: *How the Leopard Changed its Spots. The Evolution of Complexity*, New York et al. (Charles Scribner s Sons) 1994.

GRAY, R.: "Death Of the Gene: Developmental Systems Strike Back", in: P. GRIFFITHS (Ed.): *Trees of Life*, Dordrecht/Boston/London (Kluwer) 1992, pp. 165-209.

GRIFFITHS, P. E., GRAY, R. D.: "Developmental Systems And Evolutionary Explanations", *The Journal of Philosophy*, 91 (1994), pp. 277-304.

HAMPSON, R.K., ROTTMAN, F.M.: "Alternative Processing of Bovine Growth Hormone mRNA: Nonsplicing of the Final Intron Predicts a High Molecular Weight Variant of Bovine Growth Hormone", *Proceedings of the National Academy of Science*, 84 (1987), pp. 2673-2677.

HANSON, M. R.: "Protein Products of Incompletely Edited Transcripts Are Detected in Plant Mitochiondria", *The Plant Cell*, 8 (1996), pp. 1-3.

HANSON, M., SUTTON, C. A., LU, B.: "Plant Organelle Gene Expression: Altered by RNA Editing", *Trends in Plant Science*, 1 (1996), pp. 57-64.

HERBERT, A.: "RNA editing, Introns and Evolution", *TIG - Trends in Genetics*, 12 (1996), pp. 6-8.

HO, M.-W.: "On not Holding Nature Still: Evolution by Process, not by Consequence", in: M.-W. HO, S. W. FOX (Eds.): *Evolutionary Processes and Metaphors,* Chichester et al. (Wiley) 1988, pp.117-144.

HO, M.-W., FOX, S. W.: "Processes and Metaphors in Evolution", in: M.-W. HO, S. W. FOX (Eds.): *Evolutionary Processes and Metaphors,* Chichester et al. (Wiley) 1988, pp. 1-16.

KITCHER, P.: "Genes", *British Journal of Philosophy of Science,* 33 (1982), pp. 337-359.

KITCHER, P.: "Gene: Current Usages", in: E. FOX KELLER, E.A. LLOYD (Eds.): *Keywords in Evolutionary Biology,* Cambridge, Mass./London (Harvard Univ. Press) 1992, pp. 128-131.

KNIPPERS, R.: *Molekulare Genetik,* Stuttgart, New York (Thieme) [7]1997.

LEE, W., HASLINGER, A., KARIN, M., TIJAN, R.: "Activation of Transcription by Two Factors That Bind Promoter and Enhancer Sequences of the Human Metallothionein Gene and SV40", *Nature,* 325 (1987), pp. 368-372.

LEVIN, B.: *Genes*, New York et. al. (Wiley) 1985.

LEVIS, R., HAZELRIGG, T., RUBIN, G.M.: "Effects of Genomic Position on the Expression of Transduced Copies of *White* Gene of *Drosophila*", *Science*, 229 (1985), pp. 558-561.

LEWONTIN, R.: *Biology as Ideology: The Doctrine of DNA,* New York (Harper Collins) 1993.

MCCAIN, R. A.: "Critical Reflections on Sociobiology", *Review of Social Economy,* 38 (1980), pp. 123-139.

NEUMANN-HELD, E. M.: "Die moderne Biologie auf der Suche nach dem 'Heiligen Gral'" ["Modern Biology in Seach for the Holy Grail"], in: T. FISCHER, R. SEISING (Eds.): *Wissenschaft und ôffentlichkeit [Science and the Public],* Frankfurt/M. (Peter Lang) 1996, pp. 137-163.

NEUMANN-HELD, E. M.: " 'Gene' können nicht alles erklären", *Universitas*, 52 (1997), pp. 469-479.

OYAMA, S.: *The Ontogeny of Information. Developmental Systems and Evolution,* Cambridge et. al. (Cambridge University Press) 1985.

OYAMA, S.: "Evolutionary and Developmental Formation: Politics of the Boundary", in this book pp. 79-104.

PHREANER, C. G., WILLIAMS, M. A., MULLIGAN, R. M.: "Incomplete Editing of *rps12* Transcripts Results in the Synthesis of Polymorphic Polypeptides in Plant Mitochondria", *The Plant Cell,* 8 (1996), pp. 107-117.

POLLARD, J. W.: "New Genetic Mechanisms and their Implications for the Formation of New Species", in: M.-W. HO, S. W. FOX (Eds.): *Evolutionary Processes and Metaphors,* Chichester et al. (Wiley) 1988, pp. 63-84.

PORTIN, P.: "The Concept of the Gene: Short History and Present Status", *The Quarterly Review of Biology,* 68 (1993), pp. 173-223.

REHMANN-SUTTER, C.: "Was ist ein Lebewesen? - Zur philosophischen Herausforderung durch die Molekularbiologie" ["What is a Living Being? - Toward the philosophical challenge of molecular biology], *Scheidewege: Jahresschrift für skeptisches Denken,* 23 (1993/1994), pp. 142-159.

STERELNY, K., KITCHER, P.: "The Return of the Gene", *The Journal of Philosophy,* 85 (1988), pp. 339-361.

WATERS, C. K.: "Genes made Molecular", *Philosophy of Science,* 61 (1994), pp. 163-185.

WATSON, J. D., CRICK, F. H.C. (1953a): "Molecular Structure of Nucleic Acids: a Structure for Deoxyribose Nucleic Acid", *Nature,* 171 (1953a), pp. 737-738.

WATSON, J. D., CRICK, F. H.C. (1953b): "Genetical Implications of the Structure of Deoxyribonucleic Acid", *Nature,* 171 (1953b), pp. 964-967.

Part III

Redrawing the Boundaries Between the Narratives of Evolutionary Biology and the Narratives of Hermeneutics, Mythology, and Philosophy

Chapter 6

The Role of Hermeneutics in Biology

SERGEY V. CHEBANOV

SERGEY V. CHEBANOV

Foreword

This report attempts to sum up the many years of work (since 1972) of the Seminar on Theoretical Biology (now the Seminar on Biohermeneutics of the St. Petersburg League of Scientists). The Seminar deals not only with the subject of this report, but with that of the conference at hand as a whole. In addition, in Tver' in February 1996 there was a workshop on theoretical biology and theoretical economics held by the Seminar and the V.A. Naishul Private Institute for a National Model of the Economy (proceedings in preparation), where the centre of interest was the problem of semiosis and interpretation as key processes in the functioning of living beings and economics.

I. Introduction: From Biosemiotics Back to the Hermeneutics of Biology

The 20th century has been given many epithets and may also be called the century of semiotics. First, because it is marked by the explication of semiotic ideas inherent in German linguistics (e.g., H.G.K. von der Gabelentz) in Saussure's semiology and in the notions of the school of Baudouin de Courtenais and Scherba about language and speech. It is marked secondly by the cybernetic notion of signals, and thirdly, by G. Gamov's speculation on the semiotic nature of the genetic code.

The pioneers had no name for the field they discovered. Later on it was named biosemiotics, apparently by Yuri S. Stepanov (Stepanov 1971). Earlier the term "zoosemiotics" was used by T. Sebeok (Sebeok 1972). The term of phytosemiotics was also used. For whatever reason there is a domain in modern biology, which being regarded from the point of view of semiotics, is usually defined as biosemiotics (Sharov 1990). But semiotics is not the only approach to sign-like structure. A more fruitful approach is hermeneutics.

Hermeneutics (or rather *Hermeneutica Sacra*, from Hermes), goes back to Hesiod, who used it in his comments on Homer. Antiquity developed the art of interpretation and passed it onto the Christian Fathers of the Church, who then adapted and modified the art for the needs of interpreting the Scriptures.

For *Hermeneutica Sacra*, semiotic structures (symbols first) are significant parts of the world, the Scriptures being the most significant of all. Approaching the world as a "sacred text" permits the limitless penetration into the sense, each one to the extent of his own characteristics and abilities. These abilities depend on having an affective way of understanding things, the righteousness of the comprehension and finally all the details of one's way of life, including clothing, diet, etc., the latter being a kind of a means of interpretation or at least the sufficient conditions for interpretation.

Hermeneutica Sacra in Europe had completely died out by the end of the Middle Ages, but in Russia, by contrast, this discipline survived in church schools until Soviet times. In fact, the philosophy of the Silver Age in Russia was to a large extent inherited from the *Hermeneutica Sacra*, if not its continuation. At least the "Philosophy of Name" (Paris 1952) or "Philosophy of Economy" (Bulgakov 1982) by S. Bulgakov and the works of P. Florensky were undoubtedly hermeneutic. Further more, it was not accidental that hermeneutic works were published in the Soviet period, in publications on semiotics (see, for example, the "Structure of Word" by P. Florensky published in the "Context 72" - Florensky 1973). The current interest in such problems in Russia and the new work being done in this field (V. Bychkov, V. Zhivov among others) allow us to state that the tradition of *Hermeneutica Sacra* is alive in Russia even now.

Reference to many types of texts previously outside the field of studies (for example advertisements, posters, warning notices and the like, no matter how correct their language is) with the purpose of solving practical problems (efficiency of communication, advertising, speed of perception of text, correction of speech, etc.) has led to the use of diverse material from psycholinguistics, the physiology of speech, communicology, etc. in semiotic studies. The development of linguistic pragmatics is also very important. Applied linguistics has included the consideration of quasi-language phenomena in computers, information systems and living beings.

All these facts support the thesis that the development of representations about language has reached a new stage designated as pragmalinguistics (I.P. Susov from Tver', Russia). At the centre of pragmalinguistics there is language as a means of practical activity of the person, or the teleological transformation of the world by him.

Thus is appears that very many concepts of *Hermeneutica Sacra*, first of all the Alexandrian symbolic allegorical school, again find meaning. There are, however, at least two differences. Firstly, pragmalinguistics is anthropocentric (instead of as theocentric *Hermeneutica Sacra* is). Secondly, *Hermeneutica Sacra* is a system, but pragmalinguistics so far represents only a collection of concepts.

Besides *Hermeneutica Sacra* there are

1) Philological historical hermeneutics, founded as a section of philology since the beginning of the 19th century (Chlodenius, Mayer, Dilthey, Ranke, Droysen). It develops the tradition of the Antiochic historical school of *Hermeneutica Sacra*.

2) The philosophical hermeneutics of Heidegger and Gadamer came in to replace philosophical ontology.

3) The hermeneutics of everyday speech (Chebanov/Martynenko, in press; cf. V. Dahl: *The Dictionary of the Living Great Russian Language*).

4) The hermeneutics of specialized texts, non-fiction hermeneutics, engaging by interpretation scientific, political, business, advertising and other texts (non-sacred and non-fiction), and the hermeneutics of separate disciplines (biology, economics) or types of texts (biological, economic).

As a result, modern semiotic studies of language in man, considering the unity of communicative and cognitive phenomena in their broad actual context - this is what I mean by "pragmalinguistics" (Susov 1983) - comes close to the hermeneutic approach (Chebanov/Martynenko 1990b). That is why I find sufficient reasons to qualify the domain of biology I have mentioned above as biohermeneutics (in a broad sense), which includes biohermeneutics *stricto sensu* and the hermeneutics of biology (Chebanov 1993).

The basic concept for the investigation I am exposing here is "enlogue" (Chebanov 1984, 1993, 1994; Russo/Chebanov 1988). The enlogue can be interpreted as quasi-dialogue taking place in quasipersonal situations when the participants have no such rational means of communication as, for instance, natural language: take the interaction between a living being and a sensible being (i.e., biologist) - the interaction is not verbal, though obviously it generates some information in the sensible being.

II. Biohermeneutics: Semiotic Means in Living Beings

1. The Status of Biohermeneutics

Biohermeneutics studies the semiotic aspect of living being as centaur-object (after G.P. Shchedrovitsky - Chebanov 1988, Schedrovitsky 1971, Shchedrovitsky 1982, 1995).

The concept "centaur-object" introduced by G.P.Shchedrovitsky - the most prominent Soviet methodologist, the founder of the Moscow methodological circle included Merab Mamardashwili, Alexander Zinoviev (the author of *Gaping Heights* and *Katastroika* = Catastrophe + Perestroika, living now in Germany), Vasily Davydov, Vladimir Lefevr (the author of *Conflicting Structures* teaching Russian prose in USA now), Oleg Genisaretskii (one of theoretical leaders of the Russian national movement now) and others (Shchedrovitsky 1995).

The centaur-object is characterized by heterogeneity, heterohierarchity, and heterochronicity, i.e., has many semantic dimensions reflected in interdisciplinary descriptive pictures, and each of them describes it as the unity of a substratum (semantophore) and its sense.

In this light, somatic and physiological organization of living beings is functioning as *semantophore* - an exponent of the semiotic unit, in which the nature of the substratum is important for its semiotic performance.

In the terms introduced by Hjelmslev (Hjelmslev 1953), this process of semiosis is schematically represented in Fig. 1. On the left, the synthesis of a semiotic means is described (indexes 1), on the right - the process of its analysis (indexes 2).

2. Molecular Genetic Processes

The most revealing is the concept of genetic code. The very idea has a strong semiotic background.

What is crucial here is the understanding that we are dealing with is a code rather than an ordinary object (Ycas, 1969). Another crucial point is that the relation between content (C) and expression (E) is not motivated (the properties of aminoacids are not derival from the properties of nucleatide triplets) (Fig. 1, Relation 3), which testifies to the semiotic, not physical or chemical essence of the code. The t-RNA structure provides this unmotivatedness because its adapters and acceptors are combined according to historically formed norms and not according to the laws of causal conditioning (cf. artificial t-RNA with violated norms of adaptor-acceptor combination). On this basis a wide practice of operation with genetic texts is now formed, not very different from traditional lexicography (card-indexes of DNA primary sequences, corresponding to the content-substance 1, or m-RNA consequences, corresponding to the expression-substance 1).

The direct and inverse degeneration of the code, linearity of nucleotide chain, triplets - punctuation marks, etc. - all these assimilate genetic mechanism and language, while from the philological point of view the question of text homonymy arisen (though genosystematics denies the very possibility of it). On this basis the banks of nucleotide and aminoacid sequences are being created (Gene Bank, USA; EMPL, Germany). Such facts as DNA transcription in two directions, DNA with a frame shift, and the existence of untranscribed loci bring to mind concepts of palindrome, letter sequences allowing more than one division into words, and anagrams.

Some investigations introduce the study of molecular genetic mechanisms into the sphere of hermeneutics.

First, the importance of expression-substance 1 is revealed - thus, the frequency of point mutations, due to the isomorphity of content-substance 1 and expression-substance 1, is treated as a result of DNA nucleotides tautomery (Ladik 1972). The necessary level of mutability is kept by the correction of a part of mutations by the systems of reparation. So, similarly to the importance of the author's and the copyist's personalities and their relationship in hermeneutics (Averintsev 1977), the content-substance 1 and the processes of m-RNA transcription, translation and maturation turn out to be important to the semantics as they determine expression-substance 1' meaning (relations 2-4 on the Fig. 1).

Thus expression-substance 1 is not an accident or attendant but of principal and even sacral importance, as it is in *Hermeneutica Sacra*.

Secondly, the role of diversity is made clear. In biosynthesis (carried out from decoding a text to its interpretation), attention is centred on the code degeneration, frame shift, regulative elements' interaction, and splicing - all these providing for the necessary correlation of content 1 and expression 1, with one gene being able to answer for the synthesis of up to 15 products. The peptides can be interpreted even more variously (rel. 6-8) when their activity depends on the state of the cell; the cell state changes like a personal image of the world as a result of a comprehension of a text.

Such a multiplicity of interpretations requiring conceptual efforts of pragmalinguistics is a fundamental idea of *Hermeneutica Sacra*, as expressed, e.g., in the notion of four senses of the Gospel: literal historical, allegorical, moral metaphorical, and salutary anagogic, each of which has own recipients.

Third, some processes (rel. 5-9) consist in the activity of fermentative centres (expression-form 2) of related proteins (expression-substance 2), formed by tertiary and quaternary structures (rel. 5). It changes the proportion of substratum and product (content-form 2), generating the changes in physiological processes (content-substance 2) characteristic of the cell state (intracellular umwelt in Jacob von Uexkull's sense - Uexkull/Kriszat 1934). Thus, genetic texts function as performatives, the interpretation of genetic symbols being carried out inside a cell. The eczone-domain correlations revealed by Belorussian genetist O.V. Davydov (Davydov 1995), leads to the disclosure of the motivatedness of genetic symbols.

The mentioned properties of the genetic code may be interpreted without hermeneutics, but the concepts are much better coordinated with it (cf. Laboratory of Molecular Hermeneutics of University of Connecticut Health Center studies).

Moreover, some properties of the genetic code cannot be discussed without hermeneutics.

For example, nucleotide triplets in the apices of icosaheders, as found by Volokhonsky (1972) from St.Petersburg (now in Germany), demonstrate a kind of symmetry that can be discussed only on the basis of the alphabetic ideas of *Hermeneutica Sacra* or other sacral hermeneutics, such as cabbala or those of some Gnostic sects. Integer characteristics of nucleotide triplets, as shown by Kalinin (1981) and Kushev (1994) from St. Petersburg, are of a purely arithmological nature and refer to Pythagorean ideas, systematically conceptualized in *Hermeneutica Sacra*. Thus, hermeneutics covers the whole range of semiotic molecular genetics matters. This is why biohermeneutics should be seen as the most universal concept of semiotic structures in living beings.

Let us similarly, though more briefly, now look at some other biohermeneutic processes.

3. Intercellular Recognition

To this group of the processes belong immunological recognition, the penetration of a virus into a cell, of a spermatozoon into an ovum, the interaction of a hormone with a target-cell and that of a mediator and postsynaptical membrane.

In principle, the scheme is the same, but is activated by the expression-substance 2‘ - expresion-form 2‘ - content-form 2‘ - expression-substance 2‘ components secreted by a cell (rel. 5‘), while the semantophore can be modified acquiring expression-substance2‘ as its expression. Then the functional groups or domains (expression-form 2‘) are to be recognized by the correspondent receptors of the membrane, and are to change their state (content-form 2‘), so as to initiate the related process (content-substance 2‘‘) modifying the state of a cell which, in its turn, can indirectly (rel.10) influence the genome of the first cell (the epigenetic component, though important, is not taken into consideration here).

The intercellular recognition builds up intraorganism communication, using the chains of intercellular contacts (nervous system) or semantophores produced by some and transferred to other cells (hormones).

4. Interorganism Recognition

Several different situations must be distinguished here.

4.1. Rel. 5‘ corresponds to the semantophore being brought not only out of a cell but out of the organism itself. These are secondary metabolic products of microorganisms, pheromones and markers of a territory (see the concept of an ecological code, proposed by Moscow investigator Alexander Levich - Levich 1983), which, as in the case above, is recognized and used to initiate related processes.

4.2. Rel. 5‘ also reflects the forming of behavioural signals, such as sounds or gestures intended for neural perception. In this case, the expression-substance 2‘ semantophore proves to be a functional manifestation of a phenotype as a result of molecular genetic processes. The perception of expression-substance 2‘ is based on the determination of the invariants of perception, as described by Pribram and Gleser (Bekhtereva et al. 1977; Gleser 1985; Pribram 1971) - expression-form 2‘ - based on functional morphology of brain. The expression-form 2' initiates related neural processes, thus forming content-form 2' and content-substance 2'.

4.3. Interorganism recognition is the foundation of the "recognition species concept", that species differentiation in bisexual organisms is based on admissible sexual partner recognition (Paterson 1985).

4.4. The fourth situation uses as expression-substance 2', an object which is not a communicative means of an organism, but is, nevertheless, perceived by a living being as semantophor. The perception is based on finding out the invariants of perception (expression-form2'), which constitute the umwelt of the organism. The expression-form 1' is also used to construct the image of a situation (content-substance2') and its representation (content-form2'). So, all these kinds of processes are analogues to the cellular (2.2-2.3), but the former are brought out of the organism (cf. the living beings dialogue with environment - in the Moscow philosopher Shukov sense - Shukov 1991).

5. Language Communication in Man

Here content-substance 1 appears as the mental representation of an object or situation (a significat), logical structuring of which organizes content-form 1 (a concept), expressed by psychic means of expression-form 1, determined by the natural language and expression-substance1, chosen by the addressant. Then, expression-substance 2 is to be perceived by an addressee (rel. 5') to form expression-form 2 (strictly speaking, its analogue expression-form 2', which can differ from expression-form 1 in, e.g., thema-rhematic organization). Here rel. 2-4 are mental processes, 1 and 5 are processes of physiological reception (see 3.4.2), 6 is perception of visual or acoustic (on the two formantas see Jakobson et al. 1963) image, 7,8 and 9 are psycho-physiological processes, and the rest are based on physiological components (cf. brain codes - Bekhtereva et al. 1977; Gleser 1985; Pribram 1971).

6. Biohermeneutics and Economics

Interpretation processes in semiotic situations are what enable us to distinguish between the economy - no matter how widely defined - and virtually universal processes of exchange. In this case the existence of the semiotic level secures the maintenance of balance, in a way that is of a transcendental and not of an immanent nature. In metabolism, such a balance is secured by the fact that the enzymes as the regulators of metabolism are in their turn metabolites. The situation is similar to natural exchange. Different metabolic processes, being commensurable by their ATF consumption or, more widely, by redox potential change, are somewhat analogous to the convertibility of currencies (cf. Witt, in this volume).

On the other hand, all economical processes are of semiotical nature, but carried out on the ground of *Homo sapiens,* who in his turn functions as a biohermeneutic system.

All of the above allows to consider the economy as a multi-interpretational formation.

7. Morphology of Biohermeneutics

In the study of all the processes the attention has been focused mainly on expression-substance 2 and expression-form 2, and on their rel.6 (the study consisting mainly in picking out semantically relevant components of the exponent). The most striking example of it is the formanta conception of phonemes recognition, involving psycho-physiological mechanisms.

Clearly distinguishing phonemes and sounds n linguistics is very instructive for biologists (as it is at the *-etical* and *-emical* level in general). For example, it would be senseless to talk about a big or a long phoneme, despite the fact that some phonemes can be represented by long or short sounds, likewise long or short morphemes do not exist, though they are represented by long or short segments.

In the same way, biologists will not speak about a big or small genotype, while a genome can be said to be big or small. However, the situation differs when speaking about the gene. They often refer to big or small genes and nucleic sequences, ignoring the similarity between genes and Platonic ideas (Liubischev 1925).

The least studied among the processes is rel. 3, treated as correlative (the mechanism of correlation can obviously be observed in adapter-acceptor relations in t-RNA).

Rel. 1 and 2 are better investigated in terms of molecular genetic processes and, partly, as the sphere of human communication and cognition (cognitive linguistics, ethnosemantics, psycholinguistic studies of referention).

Rel. 4 is considered in detail in molecular genetic studies and, to some extent, a measure is used to describe processes in other domains (natural language, in generative phonetics). Rel. 5 attracts the attention of biologists, as well as of specialists in the theory of communication. The process of rel. 7 is always multiform and has diverse consequences. Biologists study it in all the three groups of situations, whereas philologists do so mainly phenomenologically.

There are some sporadic studies of rel. 8 and 9(10). Of particular interest is the research of the languages of the brain, potentially unifying biology with linguistics (Bekhtereva et al. 1977; Pribram 1971).

Speaking about rel. 3 and 4, the attempt to describe them in generative grammar is to be noted. But *there* these relations are considered independently of psychophysiological processes. Biologists, on the other hand, carry out close studies of the mechanisms and semantophores (inquiries in nuclein acids and their functioning in the molecular biology of the gene).

The general tendency, briefly, is for phenomenology to be described by linguists, and mechanisms by biologists. Thus, biologists, to use the terms of the Russian biosemiotician Alexei Sharov now working in USA, consider a text together with its author, the printing-press, and the reader (2.5) (Sharov 1990), while among philological disciplines only hermeneutics take all these into consideration, other branches considering the text outside its broad actual context.

In view of the structural isomorphism of the four mentioned groups of processes, we postulate their self-modelling relations: 2.3 in 2.4 and 2.5, 2.4.1 in 2.4.2 and 2.5, 2.4.2 and 2.4.3 in 2.4.5. This can be represented in the Fig. 1 as building on a new chain on the right.

Accordingly, any hermeneutic interpretation where a human being, as an interpreter, is considered, involves biohermeneutics as a constituent, thus addressing the theoretical basis of hermeneutics.

It should be noted that hermeneutics has been taking shape along with the physicalist study of the semantophore, and the idea of the existence of a *plan du contenu* has been introduced only in extreme situations, involving a leap in our cognition of the world (Levich 1983; Morozov 1978; Ycas 1969).

III. The Hermeneutics of Biology

1. Sensible Being - Living Being Enlogue

These enlogues include 5 components (Fig. 2): sensible being, living being, the enlogue itself, enlogy, and the text, where the enlogy is fixed. Every component of enlogue can be variously interpreted from different points of view - those of common thinking, mythology, religion, and also professional ones, such as those of biology, medicine, veterinary, agriculture, forestry, etc., and those of philosophy, ideology, art and engineering (especially biotechnology).

The interpretations are not only concerned with "scientific" biology, and it is natural to treat biological conceptions in a general survey in this way. Now, however, we discuss the interpretations relevant to biology - with the interaction humanities, treating sensible being, living being, enlogue and partly enlogy within the framework of Heidegger - Gadamer hermeneutics, and text - in the traditional hermeneutic understanding.

2. Ontological and Epistemological Hermeneutics of Biology

All the enlogue components can be considered in two ways: answering the question of how fit they are to reflect the nature of living beings, or the question of what the component is and why. The enlogue is regarded, respectively, in ontological (showing enlogue - living being essence relations) and epistemological (considering the nature of means to achieve a goal) hermeneutics of biology (biology in broad sense).

The ontological hermeneutics of biology has a normative value character and qualifies the components of cognitive activity concerning living beings as unacceptable (if they do not reflect the nature of living beings) or acceptable only to some extent. In scientific biology ontological hermeneutics of biology practically coincides with methodology, but there is no Midas' paradox (a change in methods caused by trying to reflect on the methods) as described by modern Russian philosopher M.A.Rozov in ontological hermeneutics of biology (Rozov 1977).

The main task is, therefore, to reveal how the reconstruction of essence can be carried out on the basis of phenomena (latent structures on the basis of explicit ones, the whole on the basis of its part, live state on the basis of fossils), how to distinguish facts from artefacts, to evaluate the biomorphness of the components of inquiry, and to establish the referents of suggestions.

The epistemological hermeneutics of biology considers any professional activity of a biologist regardless of whether it gives any adequate representation of living being's nature or not. In the absence of any normativity the task is to interpret the relation of a sensible being with a living being, be it fruitful and methodologically correct or not, whether its results be adequate or false. That is why, from this point of view, work containing many methodological faults or none, done by students or by recognized scholars, by people with mental disorders, or in various emotional states, are equally interesting.

The methodology of biology can also be interpreted by the epistemological hermeneutics of biology, and therefore, methodology must be included in the ontological hermeneutics of biology, while other components of biology are the subject matter of the epistemological hermeneutics of biology.

3. Interpretation of Living Being: Hermeneutics of Fragments of the World

Social cultural types differ in their interpretation of living beings: there are vitalist biologists, for instance, and reductionists, and there are approaches which involve a system and semiotic conceptions, other which do not, etc.

Two circumstances can be elucidated:

1) the result of interpretation depends on the point of view, and

2) more significantly, the proponents of different points of view can reach agreement on what they interpret deictically (i.e., by indicating the referent in a set of objects, such us a collection, a kitchen, or a vegetable garden).

On this is based the comparative questioning method of research on medicinal plants by scientists from St. Petersburg and Ulan-Ude (Aseyeva et al. 1985, Badaraev et al. 1982), and typification in biological taxonomy (Jeffrey 1978). In this way, it can be explained how certain units of the Universe are chosen as living beings, with little intercultural divergence in their determination as such.

4. Interpretation of Sensible Being: the Hermeneutics of Biologists

The primitive syncretic mind typically identifies a sensible being with a living being - totem attributes to a sensible being the habits of a living being, and in common thinking and folk psychology the comparisons of a person with some animal are still widespread (cf. e.g. Medieval Bestiary, 1984). Generally speaking, a sensible being is a morphologically organized interpreter, and the traces of its organization can be found in correspondent enlogy, which thus turns out to be anthropomorphic.

This anthropomorpho-generative orientation of sensible being prevailed in the Biblical, Medieval and Renaissance tradition in Europe, and within this framework a sensible being in its cognition of a living being is regarded as acting in its unique way, proceeding from its individual attitude towards the situation. As a reaction to this, there appears new Galilean-Cartesian scientific methodology, where the anthropomorphism is to be overcome by the explication of human cognitive abilities (vision, mind, simple manipulations), thus reducing a sensible being to an unstructured point subject (as opposed to a structured object), and where cognition in the boundaries of logic epistemical approach is understood as a reflection (Shchedrovitsky 1982, 1995).

In recent decades some attention has been paid to the cognitive activity of a subject, and the subject itself has been again endowed with different abilities, including the activity of sensible being, seen again as a personality Polanyi (1962). This latter view is that of epistemological hermeneutics of biology.

The concept of sensible being as the personality of the biologist requires us to accept, at least, the following components:

1. The *culture,* and first of all the *ideology*, including the professional one. Mastering one's culture by a personality implies education and bringing up as special "standardizers" of personality. Particularly important is the professional culture of a biologist (his awareness of methods and theories). In the European tradition, involving the Biblical idea of humans as sovereign, biomorphism is not proper (in contrast to Indian tradition, within which the relation of sensible being and living being is based on metapsychosis), and hylozoism, except Aristotle, Paracelsus and, say, Steiner's anthroposophy, is not presented in it, because it is orientated in Europe to substrate conceptions, and in America to functional ones. At the same time, in Abrahamite re-

ligions there is a concept of the Book of Life (Averintsev 1977), where all living beings are presented.

Cultural premises of professional activity, including biology, are considered in metabletics (historical psychology founded by van den Berg - Berg 1961), in studies in Leyden, in which family life, sex and age differences, social status and style of life are examined (Engel 1970; Jacobs 1980). In this context, inexplicable growth of population in Europe in the first quarter of XVIII century, Linne's revolution in the description of plants, and the appearance of "L'esprit des lois" by Montesquieu are juxtaposed.

Similar problems are discussed by bryologist F. Verdoorn in his biohistory, where the relations between man and living being, and the cultural reflection of these relations, are reexamined (Verdoorn 1944, 1966). The Institute of Biohistory in Utrecht investigates dogmatic, psychological and philosophical premisses of biological conceptions in different cultures, continuity and succession of biological ideas and schools, using not only treatises but folk bestiaria (Medieval Bestiary, 1984), legends and myths, and studying the etymology of taxonomic names and the symbolics of biology (Foucault 1971; Kull/Koppel 1989; Stearn 1966).

Methods of H.Garfinkel's ethnomethodology, which include, in the domain of biohistory, ethnolinguistics (ethnosemantics - cf. taxonomic investigations by Americans: Labov 1978; Lakoff 1986), are important for biologists because man's everyday life has its biological component (sleep, nutrition, reproduction), are also involved. Studies of the sacralization of living beings (particularly in the Bible - Biblical botany and zoology - Moldenke/ Moldenke 1952) and of mythical conceptions of living beings are also in the domain of biohistory. At present Dr. Tatjana Vishenskaja from our Seminar is studying Biblical Botany and is preparing a publication on this subject.

2. The *social structure* projected into a biologist is considered in metabletics and biohistory studies. There are two kinds of study heret: on the one hand, extra-scientific factors such as the specific features of mentality are investigated, and on the other, from the point of view of sociological research of science, other concepts are introduced, such as paradigms and scientific revolutions (T. Kuhn), the logic of the development of science (I. Lakatos), or, on the contrary, anarchy theory of knowledge (P.K. Feyerabend).

3. The *paradigms* (for instance, the means of organization of the "intellectual eye" in Gregory's sense - Gregory 1970) comprise some "canonized" fragments of the world picture, and all new concepts have to be made compatible with them. Some components of the paradigm can become objects of re-

flection. In some cases all norms are rejected (cf. anarchy theory of knowledge).

4. *Mentality* is seen as a set of unconscious ideas about certain realities, which exists implicitly in the mind (particularly influentiol in the expert performance of a biologist - cf. personal knowledge in the chemist Polanyi's sense - Polanyi 1962).

5. *Natural language* discussed in this respect in ethnolinguistic (ethnosemantic) studies, initiated by W.Humboldt and then continued by E. Sapir and B. Worf. Studies in folk taxonomy and reference (Labov 1978) are particularly important for understanding the personality of the systematist biologist (cf. Chebanov 1984, 1987).

6. *Ethnic belonging* of a sensible being is one of the factors defining its mentality. The ideology, forming national world picture, is also ethnically specific (cf. the very original studies of Gachev 1988).

Thus, Byzantine tradition with integrity as its highest value, having been incorporated into the Russian national mentality, leads to the appearance of such personalities in science as T.D. Lysenko and M.M. Bakhtin (observed by B.M. Gasparov, oral communication). Subjects like this, including the problem of cross-cultural interpretation (Verdoorn 1944; cf. the very interesting ethnomycological investigations of Toporov 1979), are considered in psychological, dogmatic and ethnobiological trends of biohistory.

Secondly, a sort of quasi-ethnos is being build up among biologists, with such features as fieldwork, and, therefore, being connected with seasons, the connectedness of systematists with classical languages via their terminology and nomenclature (Stearn 1966), intensive international contacts, similarities in laboratory organization, data processing, means of communication, etc., even to the extent of eating similar foods, and, therefore, methods of ethnology must be used for their description. Then we can see that the obviously mythological elements are included in the professional paradigm (e.g., the opposition "Hero/Anti-hero" is seen in the confrontation of heredity and variability, and also in that of mutations and the functioning of systems of reparation).

7. *Psychosomatic organization* of the sensible being. It comprises the specifics of motorics, personal schedule of daily activities and regimen of sleep - in other words, the peculiarities of everyday life inevitably influencing the professional activity. Thus, an inclination for strict and detailed argumentation and taxonomy is apparently linked with excessive sphincter constipation, depending, in its turn, on eating patterns and physical activities.

These kinds of connections are included in the field of metabletics, and having been studied, can be useful for the purposeful development of sensible beings who are able to do fine expert work, which we see, in a sphere of science, in the Pythagorean unions (cf. their ban on eating beans) or in gnostic sects, and, outside the sphere of science, in mystical practices. Nowadays, however, biological investigations are carried out by sensible beings who are deeply unaware of their abiological way of life. At the same time, although there have been some attempts to use man's parapsychological abilities in biologists' work, its consequences may be regarded as destructive for the culture.

Usually, the seven mentioned claims are treated by most people as disputable, and dependent on one's will, but the two following claims, as a rule, are taken for granted, without any proof, as basic by "naive" methodologists.

8. *Biological interpretation of sensible being*. Since a man as a sensible being acts in the structure of biological species Homo sapiens (G.P. Shchedrovitsky, oral communication), his biological characteristics are important for understanding him as an interpreter of a living being. First of all, the species-specific traits of reception, the adequacy of an irritator to a receptor as a condition of perception, and the mechanisms of perceptional invariants (4.2, the circle is made up) must be considered here. In addition, functional states of a sensible being are important, because of their influence on living being (e.g., cells' reaction to the state of women biologists during mitogenetic rays studies), and man's microflora, which is to be taken into account by microbiologists.

There is evidence that in *Planaria* conditioned reflexes can be passed on by feeding of "untrained" worms with "trained" ones, which is thought to be related to nucleic acids.

This permits a new approach to the following circumstances:

- scriptural narration of eating the Scroll;
- scriptural narration of the letters flying up to the heaven from burning Scroll;
- the Gnostic idea (Markos) of the Divine Truth's body as composed of Greek characters;
- food restrictions in different religions as a means of comprehending spiritual truth: food as a means of interpretation;
- popular beliefs that one can obtain an animal's qualities by eating its corresponding organs.

Thus we have some new and very important intersections of biology and hermeneutics.

9. *Psychological organization of sensible being.* An important task here is to reveal the sensible being as belonging to a certain psychological type - rational(rt) or emotional(em); extra- or intravertive, etc. - and, accordingly, to find out what kind of enlogue is preferable for this type of sensible being, and also to establish the degree of coherence between the sensible being and the living being under consideration, the biographical circumstances providing for the awareness of the sensible being in the living beings, etc. (cf. psychological and biographical trends in biohistory).

Moreover, psychological accentuations, and mental and somatic pathologies should not be disregarded, due to their role in forming the professional profile of a personality (cf.the psychoanalitic hermeneutics of P. Ricoeur - Ricoeur 1975). Thus, many systematists are careless and disorderly in their everyday life, which is compensated for in their professional activity. For revealing the true motives of one's activity, the methods of psychoanalysis can be used, and the epistemological hermeneutics of biology then can be regarded as a special kind of psychoanalysis of biologists' collective unconscious, which, in contrast with individual psychoanalysis, has not been sufficiently studied.

The above observations on the organization of sensible being resolve mainly the tasks related to ontological hermeneutics of biology, being, nevertheless, important to epistemological hermeneutics of biology in respect of sensible being projection of its organization onto living being. In this case, if sensible being is considered as an unstructured subject, the type of projected structure cannot be conjectured. But, on the other hand, an opposite danger exists - that of inability to see anything but one's own reflection, one's own properties (cf. in Persian poetry; a sage who is in love moves apart the curls of his beloved/truth to see her face, but sees only a mirror with his own reflection in it). A special enquiry into this issue allows to find sensible being's focus on to a certain type of enlogue, its activity within it, and the true motives of this activity.

5. Interpreting Enlogues: Hermeneutics of Biological Research

Contact between sensible being and living being implies different types of enlogues, be it totem, occultist practice, putting in practice an algorithmized method, or dialogue.

Central position in biology is occupied by its most specific enlogue between living being and sensible being, in which living being reacts to the very process of investigation (through biological fields, its organs of sense, active component of its behaviour). Description of this enlogue refers mainly to competence of psychology and psychological trend in biohistory. Type of enlogue is determined by the sensible being's orientation (rt, em, etc.), by a degree of affinity of a biologist and a living being under study (biomorphity), which can vary in contacts with different taxa (taxon-specific - see Chebanov 1988).

The simplest way to interpret an enlogue is to check it (as ontological hermeneutics of biological activity) against the declared technique (referring to culture and ideology as a component of sensible being). Immediately, we find the divergences between the results of ontological and epistemological hermeneutics of biology. It can be revealed, e.g., that sensible being - living being interaction was in fact the enlogue of an expert (determined by mentality, biological and psychological status of sensible being), whereas it has been described as a realization of some operational techniques.

Thus, by means of a successful act of interpretation, we arrive at a reconstruction of the ontological structure of the process of sensible being - living being interaction. In this connection, to be noted is the socially acceptable degree of difference between the process and its qualification; in other words, there exists a norm of deviation from the normal situation (an element of paradigm).

Thus, nowadays biometric standards are accepted as imperative. But while they seem to be followed (culture and ideology as components of the sensible being interpreted in ontological hermeneutics of biology), there are very few works at our disposal in which the type of distribution would have been defined, the precision index of the experiment calculated, and the necessary volume of the sample determined. Nevertheless, the works are regarded as quite correct, and the professional paradigm (interpreted in epistemological hermeneutics of biology) turns out to be more significant than a normative system of methodics; i. e. the transition from ontological hermeneutics of biology to epistemological hermeneutics of biology, transferring culture as a component of sensible being to paradigm, is obvious.

Enlogues are also highly influenced by the facilities used, popular beliefs and superstitions, biologists' customs of laboratory work, etc.

6. Interpretation of Enlogy: Biological Conceptions Hermeneutics

An enlogy is formed as an enlogue result. In a predominantly rational (rt) enlogue, where the logic-epistemic approach is accepted (Shchedrovitsky 1982, 1995), the enlogy is congenial with phenomenologists' noema.

Sensible being's attitude towards the enlogy between living being and sensible being varies according to the type of this enlogue. In syncretic folk thought this enlogy is interesting as such, while in scientific biology the tendency is to detach only the components induced by living being, attributing the rest to artefacts.

The anthropomorphization of living being in enlogue is one of the most difficult subjects of the ontological hermeneutics of biology. There are many separate questions in this area e.g.:

Σ Are reason, will, aspiration to beauty, etc. the attributes of a Living Being or not?

Σ Are choice and sequence to strategy (k- and r - strategy, "patiences", "violences" and "explerents" in Ramenskii sense etc.), concepts of resources, division of functions between different structures, mutualistic symbiosis, other similar economic and social categories metaphors, or exact concepts?

a) Interpretation of Enlogy in Biology

The enlogies formed in biological investigation vary very significantly. For the sake of illustration let us compare, e.g., the results of automatic analysis of electroencephalogram spectrum, morphology of a taxon, and empirical observation on vertebrate matrimonial behavior (Table 1). The difference is so considerable that the sensible being generators of the enlogies are disposed to qualify each other's works as senseless, although it is beyond doubt that they simply belong to different branches of biology.

b) Interpretation of non-rt Enlogue's Enlogy

In an enlogue, where *em* predominates in sensible being activity, aesthetic resources of a living being are revealed (cf. aesthetic concepts in geography - Karinskii 1990; Poljan 1978), as it expresses in folk bestiaria, poetry and painting. It can make a contribution to scientific biology, not only by replenishment of factology (cf. ethology of E. Saton-Tompson and Russian biologist and writer V. Bianki), but also promoting changes of biological paradigms (cf. role of painting for Botticelli, Dürer, Rousseau, Monet and others, and of prose for A. Gid, in revolutions in botanic description - Jacobs 1980). Development of painting (as an element of culture as component of sensible being) also influences biological drawing.

Non-rt enlogies, which are usually of no interest for biologists, often give rise to new approaches. Their eurysthical role in biology being relatively important, they cannot be regarded as biological facts, because latters always result from rt and methodologically correct enlogue. However, the present interpretations of folk and especially eastern traditional medicine in western science reveal the fruitfulness of non-rt enlogies (e.g., Aseyeva et al. 1985). It is becoming more obvious that seemingly absurd traditional conceptions are not more absurd than the modern scientific idea of a bean as a fruit of a pea (Chebanov 1987, Chebanov/ Martynenko, in press).

Not-rational comprehension is accepted in the evolution theory. For example, Darwin considers evolution not from rational, but conative positions - category sifting, selection, competition, struggle for existence are conative categories. Anarchist prince P.A. Kropotkin, objecting to this, speaks about mutual aid in nature also continuing discussion in a plane of acting. The empathic categories are used in bioethics (Oleskin 1994; Vlavianos-Arvanitis 1991; Vlavianos-Arvanitis/Oleskin 1992). In bioaesthetics, biological evolution is explained by a shift of styles (emotionally). Only nomogenetists (L.S. Berg, D.N. Sobolev, A.A. Lubischev, S.V. Meyen) have delt with rational consideration of evolution processes.

c) Auxiliary Biological Enlogies

Besides the living being - sensible being enlogue, there exist some other enlogues important for biology, considered in theoretical biology, metabiology, philosophy of biology, etc., biology being in this way connected with every element of culture.

Wide use of auxiliary enlogies brings special importance to the epistemological hermeneutics of biology, by means of which living being's characteristics, induced by different types of sensible being, become visible.

d) Biological Interpretations of a Hermeneutist Biologist
Now we shall turn to the enlogies of biologists, directed to the sensible being itself. The sensible being, in this case, is a personality deep within the culture which is to be interpreted by biologists. As a rule, a biologist can act in this field mostly as a dilettante, the most interesting results being generated by psychologists or culturologists.

The interpretation of a hermeneutist biologist supplements the characteristics of sensible being (cf. sec. 4). And this description, obviously, is a description of sensible being enlogy, formed in another enlogue, particularly those which the biologist himself is generating as sensible being.

7. Interpretation of a Biological Text: Nonfiction Hermeneutics

Biologists' enlogies are expressed in biologists' speech and in their written works (Woodger 1952), which are widely reinterpreted by other biologists.

Biology, quite independently of philology, has formed its own tradition of text interpretation. Especially revealing in this sense is the treatment of taxonomic descriptions (Chebanov 1987; Chebanov/Martynenko, in press).

Some time ago, these texts - their content and their expression - have aroused interest in different branches of linguistics: in the studies of sublanguages of limited use, in ethnosemantics, and in pragmalinguistics (i.e., all the aspects - syntax, semantics, and pragmatics are actually in the scope). This interest provokes rebirth to hermeneutics (Chebanov/Martynenko 1990 b).

Non-fiction hermeneutics resolves the problems observed in hermeneutics of biology (see sec. 3-6) by finding specific pragmatic markers in the text, in order to reconstruct the initial communicative situation in which the text has been generated. Given knowledge of the situation, the semantic structure can be reconstructed. Since ability to recognize the markers and semantic structure depends on particular features of the personality of sensible being, this process must be regarded as an enlogue.

The enlogue between sensible being and biological text differs considerably from enlogue between sensible being and living being (principal for the empiricists).

For many texts of biologists, heterosemioticity - heterogeneity of pragmatics, semantics and syntax - is specific (Chebanov/Martynenko, in press). That is why they can be studied in both ontological and epistemological hermeneutics of biology.

First type of interpretation extracts from the text information about a living being. Ideally, the text provides opportunity to find out the referent described (addressing, e.g., a nomenclature type). The same task (to an extent that the referents themselves rather than their conceptualizations are interesting) includes the revealing of referents in non-biological text or in the texts of other cultures in which living beings are described. The results of such interpretation are corrected by results of other enlogues.

For the epistemological hermeneutics of some biological texts, the literary form of the text is important, being also interesting to applied linguistic studies of professional communication, and to literary history in classical languages (Stearn 1966), which makes it possible to speak about the development of biological literature as part of the literary process (observed by famous Russian poet O. Mandelschtam in a series of his articles - Mandelschtam 1967, 1969). Here not only common features of "standard" style biological treatises attract attention, but also ideostyles of authors, whose texts are stylistically marked (Linne, Lamarck, I.H. Burkill described tropic flora of Africa with Anry Gid; N.M. Gaidukov proposed ecological description of algae), or of those who showed their worth in both biology and literature (Goethe; Chamisso, a German poet of French origin, the author of the cycle "The love and life of woman", and zoologist who travelled around the world on a Russian ship "Rurik" under Kozebu governing and described Salpa's metagenesis; lepidopterologist and writer Nabokoff; entomologist, poet and translator N.A. Holodkovski; writer and biologist V.V. Bianki, et al.).

8. Morphology of the Hermeneutics of Biology

In our understanding of biology as sensible beings' activity involving living beings and different conceptions of them, we should observe that hermeneutics of biology, in contradistinction to biohermeneutics, is less structured and more versatile. It can be explained by the multimillennial experience of sensible being - living being interaction, important to the very existence of man. That is why, together with purely methodological interpretations (ontological hermeneutics of biology) of humanitarian aspect of biology, interpretations resulting from experiences of affinity and co-naturality of humans and living beings are also valuable. Thus, there are two principles in morphologization of the hermeneutics of biology.

The first is the most important to the ontological hermeneutics of biology, and consists in morphologization of the enlogue's components (see sec. 3 - 7). The second is purely culturological, when, in the enlogue typology, besides biologists' enlogues, those of primitive thought, artists, and practioners, are distinguished. This typology reflects different types and degrees of sensible being's affinity with living being. And different components of enlogue are respectively "perfect" in different types of enlogue. In this way, the more sensible being - living being is co-natural, the more profound is the comprehension of the living being's nature. But, in this case, some problems would probably appear in forming of a text, which, as well as the enlogue, would differ from the norm. In some cases, such non-legitimate "intrusions" can reform the normative conception of enlogue (methodological revolutions in biology, cf. Jacobs 1980).

This situation is the consequence of a method of expert examination, widespread in biology due to the mentioned sensible being - living being co-naturality. The nature of this method is under consideration in the hermeneutics of biology, whereas its two types form two different branches of biology - bioaesthetics and bioethics (Vlavianos-Arvanitis/Oleskin 1992).

One of the aspects of sensible being - living being co-naturality is that to a sensible being not only laws, but also norms are inherent (Chebanov 1988), and it adds ways of interpretation (without which biohermeneutics is inconceivable) in sensible being's cognition of living beings.

IV. Form/Sense Opposition in Biology

The analysis of biohermeneutics and the hermeneutics of biology reveals that all interpretation processes of the hermeneutics of biology, including hermeneutics of biohermeneutics, are based on hermeneutic mechanisms; i.e., the two disciplines consider two sides of the same process (Chebanov 1993, Morozov 1978).

Then, the interpretation processes turn out to be inherent in all protein-nuclein organisms, including human beings. Nevertheless, biology considers mostly the organisms, their forms.

The investigation of organisms' morphology has a long tradition and provided a lot of interesting data. Within this tradition, a seemingly useful idea repeatedly appeared: that there exists some sense in organisms. At the same time, biology has at its disposal many special semiotic means for fixation of knowledge about organisms, each of these means implies a certain ideology - and thus has a definite sense.

The idea of existing sense in organism has had a great impact on the progress of biological science. And - moreover - it is impossible to conceive living being as devoid of sense, the very ontology being fundamentally alternated by semiotic idea of code when it is applied, as a heuristically valuable assumption, to fine research in the substratum of life - and when such a leap forward in this research is achieved.

Now that the world is resurrecting the global integrity on the basis of semiotic conceptions, the rebirth of hermeneutics brings to biology and economics the means to reflect this integrity.

References

ASEYEVA, T.A., K.F. BLINOVA, G.P. JAKOVLEV: *Lekarstvennye rasteniya tibetskoy meditsiny* (Medicinal Plants of Tibetan Medicine), Novosibirsk (Nauka) 1985 (in Russian).

AVERINTSEV, S.S.: *Poetica rannevizantiyskoy literatury* (Poetics of Early Byzantine Literature), Moscow (Nauka) 1977 (in Russian).

BADARAEV E.D., S.M. BATOROVA, T.A. SURKOVA: "Logicheskye osnovy me-todiki rasshifrovki tibetskich nasvaniy rasteniy i ikh identifikatsii" (Logic Bases of a Technique of Decoding of the Tibetian Names of Plants and their Identification), in: L.L. HUNDANOVA (Ed.): *Materiali po izucheniyu istochnikov traditsionnoy sistemi indo-tibetskoy meditsini* (Materials on study of the sources of traditional system of Indian-Tibetian Medicine), Novosibirsk (Nauka) 1982, pp. 27-41 (in Russian).

BEKHTEREVA, N.P., P.V. BUNDZEN, J.L. GOGOLITSIN: *"Mozgovye kody psikhicheskoy deyatelnosti"* (Brain Codes of Psychic Activity), Leningrad (Nauka) 1977 (in Russian).

BERG van den, J.H.: *"The Changing Nature of Man: Introduction to a Historical Psychology"*, New York (Nortono) 1961.

BULGAKOV, S.: *"Philosophy of Economy"*, New York (Chalidze Publication) 1982.

CHEBANOV, S. V.: "Predstavleniya o forme v estestvoznanii i osnovaniya obsh'ey morfologii" (The Comprehension on Form in Natural Sciences and the Foundations of General Morphology), in: T. OJA (Ed.): *Orgaanilise vormi teooria* (Theory of Organic Form). Abstract 10th conf. theor. biology 7-9 May 1984, Tartu (Tartu Riiklik Ulikool) 1984, pp. 25-40 (in Russian).

CHEBANOV, S.V.: *Yazik opisaniya taksonov* (Language of Taxon Description) Phd Theses 1-2, Leningrad (Leningrad State Univ.) 1987 (in Russian).

CHEBANOV, S.V.: "Theoretical Biology in Biocentrism", in: K. KULL, T. TIIVEL (Eds.): *Lectures in Theoretical Biology*, Tallinn (Valgus) 1988, pp. 159-168.

CHEBANOV, S.V.: "Biology and Humanitarian Culture: The Problem of interpretation in Bio-Hermeneutics and in the Hermeneutics of Biology", in: K. KULL, T. TIIVEL (Eds.) *Lectures in Theoretical Biology. 2nd Stage,* Tallinn (Estonian Academy of Sciences) 1993, pp. 219-248.

CHEBANOV, S.V.: "Man as participant to natural creation. Enlogue and ideas of hermeneutics in biology", *Biology Forum*, 87,1 (1994), pp. 39-48.

CHEBANOV, S V., G. YA. MARTYNENKO: "Osnovnie tipi predstavleniy o prirode yazika" (Basic Types of Conceptions of the Language Nature), *Linguistica* (1990a) (= Acta et Commentationes Universitatis Tartuensis 911), pp. 112-133, Tartu (in Russian).

CHEBANOV, S. V., G. YA. MARTYNENKO: "Idei germenevtiki v prikladnoy lingvistike" (Ideas of Hermeneutics in Applied Linguistics), *Quantitative Linguistics and Automatic Text analysis* (1990b) (= Acta et Commentationes Universitatis Tartuensis 912), pp. 92-111, Tartu (in Russian).

CHEBANOV, S.V., G. YA. MARTYNENKO: *Semiotics of Descriptive Texts*, St. Petersburg (St. Petersburg State University), in press (in Russian).

DAVIDOV, O.V.: "Problem of genetic code structure: new data and perspectives", in: B.F. POGLAZOV, B.I. KURGANOV, M.S. KRITSKY, K.L. GLADILIN (Eds.): *Evolutionary Biochemistry and Related Areas of Physicochemical Biology*, Moscow (Bach Institute of Biochemistry and ANKO) 1995, pp. 283-295.

ENGEL, N.: "A Consideration of the Old Astrological Theory", in: *Essays in Biohistory and Other Contributions to F.Verdoorn*, Utrecht (Univ. Press) 1970 (= *Regnum Vegetatile*, 71), pp. 15-18.

FLORENSKY, P.A.: "Struktura slova" (Structure of Word) *Kontekst 72. Literaturno - teoreticheskie issledovanija* (Context 72. Studies on the Theory of Literature), Moscow (Nauka) 1973, pp. 348-375 (in Russian).

FOUCAULT, M.: *Les mots et les choses* (Words and Things), Paris (Gallimard) 1971.

GACHEV, G.D.: *Natsional'nie obrazi mira* (National images of the world), Moscow (Sovetskiy Pisatel) 1988.

GLESER, V.D.: *Zrenie i mishlenie* (Sight and Thinking), Leningrad (Nauka) 1985 (in Russian).

GREGORY, R.L.: *The intelligent Eye*, London (McGrow) 1970.

HJELMSLEV, L.: *Prolegomena to a Theory of Language*, Baltimore (Waverly Press) 1953.

JACOBS, M.: "Revolutions in Plant Description" *Miscellaneous Papers*, 19 (1980), pp. 155-181.

JAKOBSON, R., C.G.M. FANT, M. HALLE: *Preliminaries to Speech Analysis*, Cambridge (Mass.) (M.I.T. Press) 1963.

JEFFREY C.: *Biological Nomenclature*, London (Arnold) 1978.

KALININ, O.M.: "Geneticheskiy kod, relyativistskaya simmetriya i tablitsa Mendeleeva" (A Genetic Code, Relativistic Symmetry and Mendeleev's Table) *Tezisy dokladov Vsesoyuznoy Konferentsii "Teoria klassifikatsij i analiz dannykh"*, 5-7 maya (Abstr. All-Union Conf. "Theory Classification and Data Analysis" 5-7 May) (chast II), Novosibirsk (AN SSSR) 1981, pp. 73-75 (in Russian).

KARINSKIY, S.S.: *Synthesis of Image of a Place as a Problem of Geographical Land's Study* (Sintez obraza mesta kak problema geograficheskogo stranovedeniya). Phd Thesis, Moscow (Moscow State Univ.) 1990 (in Russian).

KULL, K., A. KOPPEL: "Elusolendite rahvuskeelesti himedest" (National names of biological taxa), in: *Eesti loodusenn rijate seltsi a astaraamat 71 kode* (Yearbook of the Estonian Naturalist's Society vol.71), Tallinn (Valgus) 1989, pp. 184-193 (in Estonian).

KUSHEV, V.V.: *730 shagov* (730 steps), St. Petersburg (Severo-Zapad) 1994.

LABOV, W.: "Denotational structure", in D. FARKAS et al. (Eds.): *Paper from Parasession on the Lexicon*, Chicago (Chicago Linguistic Society) 1978, pp. 220-260.

LADIK J.: *Quantenbiochimie fur Chemiker und Biologen*, Budapest (Akademiai Kiado) 1972.

LAKOFF G.: "Classifiers as a Reflection of Mind", *Typological Studies in Language*, 7 (1986), pp.13-52.

LEVICH A.P.: "Semioticheskie strukturi v ekologii ili sush'estvuet li ekologicheskiy kod?" (Semiotic Structures in Ecology or whether there is an Ecological Code?), in V.D. FEDOROV (Ed.): *Chelovek i biosfera* (Man and Biosphera).

Vip. 8 (Ekologicheskii prognoz - Ecological prognosis), Moscow (Moscow State Univ.) 1983, pp. 68-77.

LUBISCHEV A.A.: *O prirode nasledstvennih factorov* (On the Nature of Hereditary Factors) Izvestija Biologicheskogo NII pri Permskom universitete. T. 4. Prilozshenie (News of Biological Scientific Research Institute of the University of Perm'. Vol. 4. Appendix) 1925 (in Russian)

MANDELSCHTAM, O.: "Lamarck", in: O. MANDELSCHTAM: *Collected works*. Vol. 1. *Poetry*, Washington (International Literary Assotiation) 1967, p. 177 (in Russian).

MANDELSCHTAM, O.: "Zapisnie knizhki. Vokrug naturalistov. Pallas. Literaturniy stil' Darvina" (Notebooks. Around of Naturalists. Pallas. Darwin's Literary Style), in: O. MANDELSCHTAM: *Collected works*. Vol. 3. *Essays. Letters*, Washington (International Literary Assotiation) 1969, pp. 133-140, 160-167, 169-178 (in Russian).

MARTYNENKO, G.Ya., S.V. CHEBANOV: "Transkul'turniy perevod v spetsial'noy kommunikatsii" (Transcultural Translation in Special Communication), in: *Etnopsikhologicheskie aspekty rechevogo obscheniya* (Ethnopsychological aspects of speech communication). Abstr. workshop. Pt. 1. Moscow, Samarkand (Institut jasikoznania AN SSSR) 1990, pp.70-71 (in Russian).

MEDIEVAL BESTIARY, Moscow (Iskusstvo Art Publishers) 1984.

MOLDENKE, H.N., A.L. MOLDENKE: *Plants of Bible*, Walthan (Chronica botanica Company) 1952.

MOROZOV, A.: "Samopoznanie simpoziuma" (Symposium's Self-knowledge), *Znanie - Sila* (Knowledge is power), 10 (1978), pp. 39-41 (in Russian).

OLESKIN, A.V.: "Understanding bios through ourselves. New concepts of life", *Biology Forum*, 87,1 (1994), pp. 105-111.

PATERSON, H.T.H.: "The recognition concept of species", in: E.S. VRBA (Ed.): *Species and Speciation*, Pretoria (Transvaal Museum) 1985, pp. 21-29.

POLANYI, M.: *Personal Knowledge*, Chicago (Univ. of Chicago Press) 1962.

POLJAN, P.M.: "Geografiya i vdohnovennie resursi prirodi" (Geography and Inspired Resources of Nature), *Priroda* (Nature), 3 (1978), pp. 51-63 (in Russian).

PRIBRAM, K.H.: *Languages of the Brain*, New York /Englewood Cliffs (Prentice Hall) 1971.

RICOEUR, P.: "Phenomenology and Hermeneutics", *General Linguistics*, 9, (1975), pp. 82-102.

ROZOV, M.A.: *Problemi empiricheskogo analiza nauchnih znaniy* (Problems of the Empirical Analysis of Scientific Knowledge), Novosibirsk (Nauka) 1977 (in Russian).

RUSSO, G.V., S.V. CHEBANOV: "Forma, sterezis i enlogiya kristallov" (Form, steresis and enlogy of crystals), in: D.P. GRYGORJEV, N.P. YUSHKIN (Eds.): *Teo-*

riya mineralogii (Theory of Minerology), Leningrad (Nauka) 1988, pp. 47-51 (in Russian).

SEBEOK, T.A.: *Perspectives in Zoosemiotics,* The Hague et al. (Mouton) 1972.

SHAROV, A. A.: "Zimnyaya shkola po biosemiotike" (Winterschool on Biosemiotics), *Zurnal obschey biologii* (Jornal of General Biology), 51, 2 (1990), pp. 283-285 (in Russian).

STEARN, W.T.: *Botanical Latin,* London (Nelson) 1966.

SCHEDROVITSKY, G.P.: "Configuration as a method of construction of complex knowledge", *Systematica,* 8, 4 (1971).

SHCHEDROVITSKY, G.P.: "Methodological organization of systems-structural reseach and development: principles and general scheme", *General Systems,* XXVII (1982).

SHCHEDROVITSKY, G.P.: *Izbrannie trudi* (Selected Works), Moscow (Shkola Kulturnoi Politiki) 1995.

SHUKOV, V.A.: "Obraz biologicheskoy real'nosti kak ishodnoe osnovanie filosofii biologii" (Image of a Biological Reality as the Initial Basis of Philosophy of Biology), in: I.K. LYSEEV (Ed.): *Priroda biologicheskogo poznaniya* (The Nature of Biological Knowledge), Moscow (AN SSSR) 1991, pp. 55-69 (in Russian).

STEPANOV, YU.S.: *Semiotika* (Semiotics), Moscow (Nauka) 1971 (in Russian).

SUSOV, I.P.: "K predmetu pragmalingvistiki" (To the Subject of Pragmalinguistics), in: *Soderzatelnye aspecty predlozeniya i teksta* (Content aspects of sentence and text), Kalinin (Kalininskii University Press) 1983, pp. 3-15 (in Russian).

TOPOROV, V.N.: "Semantika mifologicheskih predstavleniy o gribah" (Semantics of mythological conceptions of mushrooms), in: *Balcanica. Linguisticheskie issledovaniya* (Balcanica. Linguistic Studies), Moscow (Nauka) 1979, pp. 234-298 (in Russian).

UEXKULL, J.V., KRISZAT, G.: *Streifzüge durch die Umwelten von Tieren und Menschen. Ein Bilderbuch unsichtbarer Welten* (Travels through the Umwelts of Animals and Man. A picture book of invisible worlds), Berlin (Springer) 1934.

VERDOORN, F.: "On the Aims and Methods of Biological History and Biography with some Notes for Collaborators of Index Botanicorum", *Chronica Botanicorum,* 8, 4 (1944) pp. 427-448.

VERDOORN, F.: "From Botanical Biography towards Animal Ecology" *Acta Bot. Netherlands,* 15, 1 (1966) pp.81-94.

VLAVIANOS-ARVANITIS, A.: *Biopolitics. The Bios Theory.* Athens (B.I.O.) 1991.

VLAVIANOS-ARVANITIS, A., A.V. OLESKIN: *Biopolitics. The Bio-environment. Bio-Syllabus.* Athens (B.I.O.) 1992.

VOLOKHONSKY, A.G.: "O formal'noy strukture geneticheskogo koda" (On formal structure of genetic code) *Tsytologija i genetika* (Cytology and Genetics), 6, 6, Novosibirsk (1972), pp. 487-492.

WITT, U.: "Evolutionary Economic and Evolutionary Biology", in this volume.

WOODGER, J.H.: *Biology and Language*, Cambridge (Cambridge Univ. Press) 1952.

YCAS, M.: *The Biological Code*, Amsterdam-London (North-Holland Publishing Company) 1969.

Figure 1

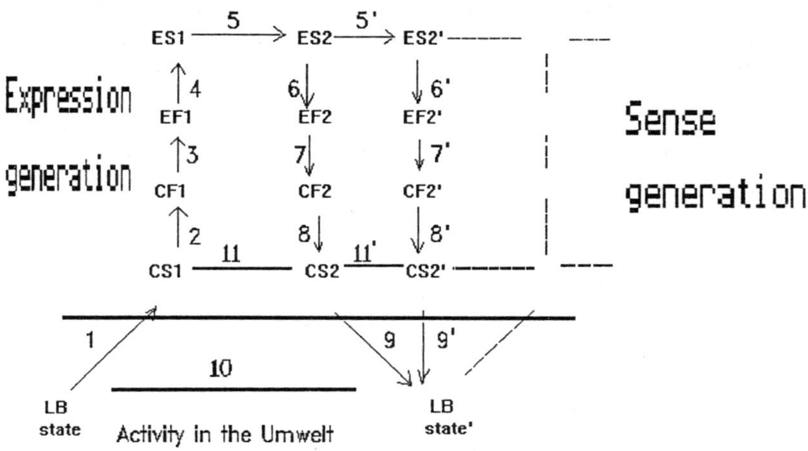

Figure 2

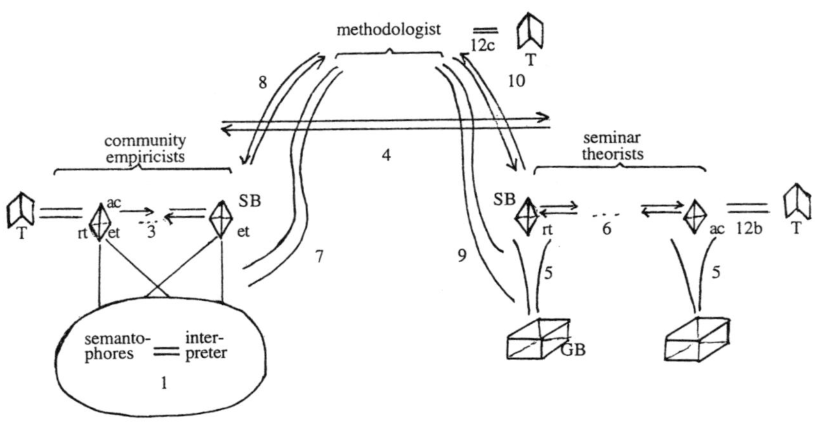

Table 1: Characteristics of biological enlogies

Abbreviations: * With much of comparative material in background, and with minimized SB influence; rt - rational, ac - active, em - emotional, et - empathetic

ENLOGY CHARACTERISTICS	EEG spectrum	Morphology of a taxon	Matrimonial behaviour
Sensible Being	subject modelling device, expanding the spectrum of perception	guild formed profile of personality	professionally accentuated personality
Enlogue	Algorythmized procedure	Examination by expert	Sensible observation *
Methodological approach	reductionism	typological approach	panpsychism
Standard discipline	physics	descriptive biology	psychology
SB biomorphity	irrelevant	taxone-specific	relevant
haracteristics of enlogy	quantitative	quantitative qualitative	qualitative
Means of presentation	numbers, diagrams	schemes, drawings	words
SB's abilities	rt	rt, ac, em, et	em, et

Chapter 7

Science and Mythology: a New "metaxu"?

JACQUES VAUTHIER

Human intelligence is concerned with what the Scholastics called "adequatio cum re". Concepts are elaborated by reality seizing. What is the counterpart of this Human activity when one has to cope with the three ultimate realities of human life: "Where do we come from?", "What is the ultimate goal of one's life?" and "Why is there is something instead of nothing?" Is it not in order to answer such essential and perennial questions and to avoid proceeding along the steep slopes of metaphysics towards the discovery of the ultimate Reality, the First Being, that mythologies or achetypes are built up? These fundamental questions pervade one's reflexion and cannot but reappear within Science. After the dramatic divorce between Philosophy and Science during the period of the "Naturphilosophie", although some philosophers have been concerned with the interrogations of the scientific world, the contrary is certainly not true. Even Wittgenstein maintains that, "Philosophy is a synopsis of trivialities" and Heisenberg, one of the founders of Quantum Mechanics affirms that, "Nothing can be done in Philosophy nowadays without reference to modern Physics."

It may therefore appear paradoxical to link Greek mythology to modern scientific knowledge which is (at the very least!) extraordinary by the pertinence of the world description it offers. Nonetheless, commentaries of scientists offering their own vision of the world, from the results obtained in their laboratories, leave the external observer in deep perplexity. Each scientist gives his or her own cosmogenesis in the light of the results of their very field of research. One discovers here a new mythology driven by scientific vulgarisation because in order to explain hard and abstract results scientists have somehow to modify the language they use between them. Precision is reserved for specialists. This dichotomy is not obvious to the man in the street who has to rely on what some scientists choose to say about what they do. Besides the difficulty of vulgarisation, an imperceptible vision of the world is injected into each explanation given. What is surprising is that the result is a sort of copy of the Greek cosmogony by Hesiodos. Now I would like to draw your attention to the common features of the two mythologies.

I. Creation and Mythology

"Before everything Chaos was, then appeared Gaïa the earth with its large womb, sturdily seated offered forever to all immortals, lord of the summit of snowy Mount Olympus, with misty Tartarus living in the depths of the earth, while Love, the most attractive of the immortal gods, in the breast of each god as well as each human being subdues the heart and the sensible will,“ explains Hesiodos in the first verses of his Theogony. Chaos, the "apeiron", the primordial empty space (from which Gaïa, the earth appeared giving birth to Ouranos the starry sky and to all the reality which is around us) appears as a plenitude expressing a density of energy that Chaos contains within itself. Let us consider what Trinh Xhuan Thuan, Professor at the University of Maryland, asserts:

"Quantum fluctuation allows the Universe to burst out spontaneously from the empty space [...]. Apparition of the Universe by this magical quantum fluctuation seems to need neither any first cause nor existing God. Its emergence can be explained only by purely physical processes." What are the processes involved within this pregnant vacuum of the Universe ? "It is not inert, it contains all the possible particles - potentially of course -, "explain Brout, Englert, Gunzig and Nardone, all four known specialists in theoretical Physics.“ Endlessly these particles burst out of vacuum state to disappear at once. Quantum fluctuations are not indeed submitted to the conservation of energy but their actualisation, the transformation of virtual particles into real particles needs an energy equivalent to the mass of the particles. It is why

Nobel Prize winner Prigogine adds “in our tepid Universe we observe only materialisation of the lightest particles [...]. The birth of the Universe would correspond with an instability which takes us from a void universe to an exponentially expanding universe thronged with mini black holes. The death of these mini black holes may have lacerated the space-time tissue corresponding to the vacuum.”

By what mechanisms, can these entire universes burst out from nothing? We have to make a small incursion into quantum mechanics, a theory which has deeply modified our understanding of the world, as well as Einstein's theory of relativity. Quantum mechanics appears at first sight as a dictionary which allows a transposition in an infinite dimensional geometry mechanisms of interaction between matter and an observer. This intellectual process is familiar to the mathematician who transfers one situation into another through the edification of a structure to help him understand the underlying reason for it and so solve the problem. Within this context, the detection of a particle, the determination of its position or its speed will be described by a mathematical operator in a geometrical space of infinite dimension. Heisenberg, one of the founding fathers of this theory, showed in his famous “relations of uncertainty”, (where the word “uncertainty” should be replaced by “lack of precision”) that the perfect knowledge of both the position and the speed of a particle is impossible. Any gain on one side imposes an increasing of the uncertainty on the other side. The variables “position” and “speed” are called conjugate variables. There exists other conjugate variables as , for example, the time-energy couple. As $E = mc^2$ where Einstein tells us, m is the mass of the particle and c the light speed, a universal constant. Every fluctuation of energy is therfore linked with a variation of mass. However any fluctuation on the time reacts on the energy and therefore on the mass which can spring out spontaneously from the vacuum. The philosophical mistake is to pretend that if an interaction is not measured exactly it cannot take place exactly. As Stanley Jaki, Templeton Prize, pointed out, the word ‘exactly’ is used first in an operational context then in a purely ontological one. For quantum mechanics enthusiasts, it is thus the measure that creates reality! By an overturning of the Greek myth, Chronos is now both generating and swallowing Gaïa and Ouranos.

Chronos, the most dreaded of their children, will castrate Ouranos his father and push him away in an everlasting expansion. On the way, the Erinnyes, goddesses of revenge will “appear along the years” Hesiodos tells us together with Aphrodite "crowned with violets" to recall the Homerical expression.

Chaos also brings a new oxymoron. This word has a special echo which is quite the opposite of that with which the mathematical theory is concerned. This theory of ultimate determinism is supposed to generate by small perturbations, almost everything from the Universe to individual free will! The theory originates from the research of Henri Poincaré, a French mathematician of the beginning of the century. The problem was to understand meteorogical fluctuations. Poincare pointed out that non linear equations may lead to new phenomena. A small perturbation in the initial data may give rise to fantastic differences further on. The beating of the wings of a butterfly in Sumatra may produce a typhoon on Nagasaki. Following the work of Gaston Julia, a French mathematician, Benoit Mandelbrojt with his new theory of fractals is now bringing forward new insights into the complexity of these equations. The use of computers lends an interesting aspect : dynamical systems - as they are called - are creating a new domain in mathematical science mixing both the conceptual approach and heavy computational methodology. Within this context, one may pose the question of computational errors. The amazing fact is that mathematicians have shown in a theorem that the accumulation of errors at each step of the calculus does not modify the conceptual results! Once more, apart from those who are involved in the subject, chaos and randomness appear linked in the same intellectual process. Nothing is more antithetical than the first word from random, for within mathematics the word random is not a synonym for a lack of understanding.

II. Life and Mythology

Once Chaos had created Gaïa, the matter, in a silent Big-Bang, the space Ouranos unfolds. Hawking, the media-prone professor of Cambridge University in "A brief history of Time" offers another version of the facts. This allows his colleague Carl Sagan of Cornell University, to write in the foreword : "This is also a book about God. Or perhaps about the absence of God. The word God fills these pages. Hawking embarks on the quest to answer Einstein's famous question about whether God had any choice in creating the universe. Hawking is attempting, as he explicitly states, to understand the mind of God. This makes all the more unexpected the conclusion of the effort, at least so far : a universe with no edge in space, no beginning or end in time, and nothing for a Creator to do." Hawking is the only one I know to try to cancel by a mathematical methodology what could be seen as the initial singularity. More classical, Weisskopf, one of the fathers of the American nuclear bomb, proposes the following scenario: "The history of the universe was undoubtedly started by an accumulation of protons, neutrons and electrons of high energy compressed inside a young star. Time of small differentiation. Later on, elementary particles aggregated to atomic nucleus and atoms appeared in the coldest regions of the star, first step towards organisation. New properties started, movement and radiation were no longer uniform. New categories of objects appeared. On the surface of colder stars and planets, temperature kept on decreasing and conditions more appropriate to the formation of a great variety of chemical components settled. But no kind of life whatsoever. Finally, in certain places of the Universe where the conditions were more suitable the great adventure of life of which we are part took place. The organic macromolecules started their cycle of reproduction and the evolution towards multiple forms of life took its way. Human life, human thoughts are but a manifestation of this phase." Unfortunately all these are but "fairy tales for grown- ups", to quote Pierre-Paul Grassé, one of the leading biologists of this century. Certainly, Miller's experiences would not convince anybody of the actual possibility of creating life from a mixture of molecules!

And Chronos? Shaken by the theory of relativity where space and time are submitted to the observer (accelerate, you will live to be older!), Chronos has no longer the impassibility of a God but his revenge is to destroy theology. Trinh Xuan Thuan shows us that this implies that God should be beyond the course of time, "But", he says, "this also raises difficulties because such a God, distant and impersonal, would not be able to secure us [...] why would he care for the progress of the human struggle against harm? The result is known to him from the beginning. Such a God out of the time would not be able to think because thinking is also a temporal activity. Knowledge would not change during the course of time. God would have to know all the changes in function of the time of any atom in the Universe [...]. Time being elastic does not allow a God to be both personal and omnipresent."

Hesiodos recounts that the Titans, Chronos and Rheia gave birth to Demeter, Hera, Hades, Poseidon and Zeus. Muses and gods live on the earth but so do Thanatos and Hypnos, death and sleep, and Oceanos, father of all the rivers. Zeus in the end will defeat his knavish father Chronos who is then ruled by this political god. The Greeks interpret sexuality as the expression of a new fecundity: Titans join together not as Chronos and Gaïa who created new beings. Apparition of life and then of human beings is in the myth of Prometheos. The descent from the golden age to the iron one is a slow fall in Ubris, the unbounded state in the presence of Dike, justice. Are we not here facing the struggle of randomness and necessity of Heraclitus reappearing in all the neo-darwinian writings? J. Monod, Nobel prize in medicine, concludes his writing with: "Man knows at least that he is alone in the unconcerned immensity of the Universe into which he merged by chance."

Deification of Chance is a new paradygm of our days. For the Latin world, chance was divided into *sors, fortuna, fors, casus.* Paradoxically enough, within quantum mechanics, chance or randomness is part of the inderminacy of matter until an experiment is made. The probability that an electron is at two places at once is not zero. This suggests at least that contemporary Physics is able to recognize its own limits.

III. Humanity and Mythology

Prometheus, friend of mankind, during a sacrifice hides the best part for the human beings. Zeus discovering the trick deprives them of fire. You know the story : Prometheus steals the fire from Mount Olympus promising humanity everlasting progress thanks to the aquisition of a technical skill. Zeus' punishment is to send a "nice harm" *kalon-kakon*, so pernicious that: "They will delight in surrounding with love their own misery," said Hesiodos. "Anthropos" they were, "andres" they will become. They find a kind of substitution of immortality in procreation, for this "nice harm" is the Woman, Pandora as Hermes calls her. It is worth quoting that for Jacquart, a biologist in France,

"Sexuality has created death: I am, therefore my parents have to die!"

Is it possible that submitting to random thermodynamics governed by the arrow of irreversibility and by bifurcations born in the chaos theory, be a version of creative evolution operating by loss and success ? From the quarks made up in the Big-Bang, man, this "auto-chtone", born from out the earth. If so we could be facing a new myth, a total inversion of the Promethean one.

At the very moment when Achilles and Hector are to fight against each other in Homer's book XXII of the Illiad, Zeus takes his golden balance and asks the two Keres, goddesses of death to weigh the souls of the two warriors. The soul of Hector is the heaviest and flies off to the Styx. What kind of eschatology does modern science offer? For the tenants of a quantum mechanics theory which rules everything, all beings will return in the everlasting cycle of molecules. Dust of stars we are, soup of quarks we will return to, dissolving ourselves into the "Great All" of the New Age. Little wonder that the oriental mystics fascinated the founders of modern physics. To speak of only two of the founding fathers of quantum mechanics, Bohr had for a coat of arms Ying and Yang, while Schrödinger gave all his philosophical explanations in the language of the Upanishads. Here too, it is interesting to see the emergence of one of the oldest myths of Humanity: the great year. All civilisations before the Judeo-Christian one like the Azteques, Mayas, Greek and Chinese relied on a circular vision of the Universe. At some unspecified time all will return to its beginnings and everything will start again. Perhaps not surprisingly this conception of the world nipped in the bud any attempt to build a real science. Amazingly enough when science is so strong, the same obsession pervades the scientific world, starting with the Big-Bang Big- Crunch theory for which there is many an incense-bearer!

What about *post mortem* life? Frank Tipler in the States proposes a "Physics of immortality". For him, the Judeo-Christian revelation on resurrection is but an avatar of Physics together with the technology of great computers. We are merely a set of informations:

"A human being has some 100,000 active genes, which means that the human genome codes some 10 to the power of one million - which is to say 1 followed by one million of zeros - humans genetically distinct. Furthermore, a human brain can contain between 10 to the power of 10 and 10 to the power of 17 bits of information which means that there are between 2 to the power of 10^{10} and 2 to the power of 10^{17} human memories possible. On this basis, there is $10^{10^6} \times 2^{10^{17}} \cong 10^{10^{17}}$ human states possible [...] which requires at most $10^{10^{123}}$ bits of computer memory: this will be available in the future."

Such a computer will be able to simulate any human being not only very well but in a perfect way. Tipler says that this computerized clone will be exactly the starting being. The information must be conserved and this is possible because we emit quantum fields which in memory have the initial data. One can ask who will make the super computer start this virtual resurrection?

Tipler's answer is: "The omega point." Is this God ? No! It is the final singularity into which the Universe will fall after the expansion we are now experiencing. This point will contain all the information and will be omniscient; one condition must be fulfilled nevertherless. Life as we know it needs to pervade all the Universe in order to be reproduced. This requires that this information be sent both by rockets moving at the speed of light minus epsilonn and auto-reproducible computers ...Besides this, the omega point will be a perfect Turing machine and therefore an omnipotent computer itself.

IV. Science, Knowledge, and Faith

This new scientific gnose already has its avatars called New Age, Deep Ecology with its adepts and priests. Feyerabend in his book *Against the Method* goes on to say that: "Science is the most recent, the most aggressive and the most dogmatic of all the religious institutions". Humanity is always reluctant to face reality which demands a struggle. No wonder that it creates its own explanation which after being locked into a golden prison, suffocates from lack of air. "*Quasi quicquam infelicius sit homine cui sua figmenta dominentur* !" says Pliny. This does not mean that Science has to be rejected, but it does raise many fundamental questions, one of which being why Human beings are able to describe the Universe with such accuracy. The problem of the true meaning of the explanation associated with abstract models formulated before any experimental verification, together with the amazing question of the intuition of the theories hereupon created, are fundamental.

No scientist can help asking himself about the miraculous fact that he is able to say something coherent about the surrounding world. This fact is so pregnant that all of them makes each day an act of faith. The Nobel prize winner Planck wrote: "Science demands a spirit who accepts to believe. Anybody who has been involved seriously in a scientific work of any kind realizes that above the door of the temple of Science is written ,We must have faith!‘ This is a quality no scientist can dispense with.” One may think that these words from Planck are lonely in the world of Science. Not at all! Heisenberg, one of the mentors of the Copenhagen School, insisted on emphasizing the role of faith throughout the work in Science: "Our driving thread is and has always been faith... I believe in order to be able to act. I act to understand. This is applicable not only to any great expeditions in a remote place but to all Science in Europe."

In the same mood, was Oppenheimer, specialist in nuclear energy and once head of the American nuclear bomb program, who said:

“It is impossible for us to make any decisive progress in the field of elementary particles research without having any faith that in this particular hetic domain of human experimentation, there is a prevailing and constraining order. One has to trust into an external universe, an order pervading all nature, into the possibility to describe this order with mathematical tools in a non equivocal way.“

He has to trust also that the Universe is homogeneous with respect to the laws he discovers and that there is a continuity even if he is not experimenting it. One striking declaration is Einstein's in a letter to Solovine. He wrote:

“One should expect the world to be chaotic, not to be grasped by thought in any way. One might (and indeed one should) expect that the world evidenced itself as lawful only so far as we grasp it in an orderly fashion. This would be a sort of order like the alphabetical order of words. On the other hand the kind of order created, for example, by Newton's gravitation theory are posited by man, the success of such a procedure supposes in the objective world a high degree of order, which we are in no way entitled to expect a priori. Therein lies the ‘miracle’ which becomes more and more evident as our knowledge develops. [...] And there is the weak point of positivists and professional atheists, who feel happy because they think that they have preempted not only the world of the divine but also of the miraculous. I have added the last point explicitly, lest you think that weakened by age I have fallen into the hands of priests.”

Of course, not all the Scientists share this vision of the world. Nobel prize winner Weinberg said that:

V. Conclusion

"Laws of Nature are not fixed once for all but are revised from time to time by a committee of dead physicists in Hell... More seriously, laws of Nature are discovered by human beings and it may not be possible to divorce the content of these laws from the psychology of their discoverers. Or perhaps there is absolutely no logical order to Nature at all?" When a physicist seriously entertains the possibility of a physical universe without order, he is exactly destroying what is he has been doing. Not surprisingly, Weinberg concludes that the universe is meaningless and blinded him to the fact that no amount of science can justify such a conclusion!

Harmony, simplicity, the possibility of expressing a law in physics using abstract mathematics, famous optimum of nature strategies as observed by Fermat, Leibniz and Maupertuis, are all extra-scientific facts that everybody who is involved in Science discovery is obliged to consider. No wonder that metaphysics is present at each corner of Science.

In Hesiodos' Theogony, it is only when evil escapes into the world that Pandora's jar reveals its precious, hidden secret. Pressed against the inner rim of the vase, straining to be released, is ... Hope!

References

HAWKING, S.: *A brief history of time*, London (Bantam) 1989.
JAKI, S.: *Science and Creation*, Edinburgh (Scottish Academic Press) 1986.
TIPLER, F.: *Physics of immortality*, New York (Doubleday) 1994.
TRINH, XUAN THUAN: *La mélodie secrète*, Paris (Fayard) 1988.
VAUTHIER, J.: *La Science entre image at réalité*, Paris (ESKA) 1990.
WEISSKOPF, V.: *La rèvolution des quanta*, Paris (Hachette) 1990.

Chapter 8

What Is the Impact of a Philosophically Based Criticism of Socio-Biology on the Scientific[1] Community?

ALOYS HÜTTERMANN

Introduction: How to Treat Scientific Results

The topic of my lecture deals basically with the question: How to treat scientific results? As you know, there are different ways to do this. One is the example given by Copernicus when he found that the geocentric model of universe was wrong and had to be exchanged by the heliocentric one. He decided, for whatever reason, not to get his results published during his lifetime. One can speculate about the reasons he had for this, the one given by him is: "If the great and mighty God allows such a long time to pass until a human being finds out the great design of his universe he was so great to give us, so a few years more or less will not make a big difference."

1 In this paper the word science and scientific are used in the Anglo-Saxonian meaning, i.e. natural sciences (in German: *Naturwissenschaften*).

Of course, in the Anglo-Saxon world of science we know that this way of dealing with scientific results wouldn't work anymore, and that we are supposed to come up with our results in a much faster way. The usual ways of passing scientific information through today are papers, books, congresses, symposia like this one. If you come up with new paradigms, new ideas, new results, the usual way is simply to distribute it and to wait until the scientific community accepts it. This may take quite a long time, especially in the case of rather new discoveries or theories, it may take decades or even longer. Typical examples for such long lag times are the quantum theory of Max Planck and the "jumping genes" of Barbara McLintock.

In certain cases you might not want to wait for this long process of acceptance which usually takes place because you think that it may be too long, that you are loosing too much time.

I. Waldsterben-Research at Göttingen

An example for such case, where it was felt that a concerned scientist has to do more than "only" just publishing, is the *Waldsterben*-issue. This was of course one of the hottest controversies which I fought through in my scientific career.

In Göttingen Prof. Ulrich had started in the sixties already to measure the ion-balance of forest ecological systems and he found out that we had a huge input of acid into our soils and he was very much concerned about what this input would do. In the early eighties we started with our research on the tree physiological part on it and so we came up with a very clear chain of events, 1. soil acidification, 2. weathering of minerals, 3. liberation of aluminium, ions, 4. inhibition of magnesium uptake by the roots, which means magnesium deficiency in the needles 5. root-decay followed by 6. needle shedding and 7. tree-decay. This is still so far the only theory which can explain the main symptoms of forest decline: the magnesium deficiency in the needles of conifers and their needle shedding. Another factor deposited in the environment which was found to be detrimental to trees was the lead from the car exhausts.

When we were learning about the physiological and biochemical processes leading to forest decline, we got concerned whether we would be able to stop these detrimental processes. One of the first graduate student working on this topic once even came into my office and said: "Its fine that we learn about all these mechanisms but still there is the danger that all what we do leads only to the stage that we will be able to explain our children what happened to the forests"

In this situation, we heeded the advice of the great Göttingen Professor of German, Prof. Albrecht Schöne, who said once in a complete different connection: "It will not help if you walk with your bare feet over deep water without sinking, if nobody knows about it."

So we decided to follow a two-pronged strategy. On the one side, we advised our associates, at one time we had more than thirty people altogether involved in this research in our two Departments, to employ solid research on all aspects of this complex phenomenon.[2] The first international publication which was really accepted[3] was one of the most controversially discussed paper on the Waldsterben issue altogether. The International Electricity Industry tried in vain to disprove if not our data, then at least our conclusions. This went so far that they demanded from us the original laboratory journals to check our statistics. The PNAS - paper has in the meantime developed to a citation classics in Tree Physiology, with almost hundred quotations recorded in the Science Citation Index by now.

This was the one part of our efforts. The other part was that we used any outlet which we could get hold on to pass a message to the public: all kinds of media, lectures to all groups which let us speak to them. We knew, if we did not stop both acidification of the soils and start its sanitation, then our trees would die much faster then without any countermeasures, and for getting such actions enacted, we had to inform the public as fast and as extensive as

2 Our institute has published so far on this very topic 35 papers in reviewed journals, 26 chapters in books, and four edited books. Altogether three Habilitationsschriften (D.Sc. - Thesis), eleven Ph.-D.-thesis, and thirteen Diplomarbeiten (Msc.-thesis) were conducted on this very field. The most important publications are: A. HÜTTERMANN/B. ULRICH: "Solid phase solution root interactions in soils subjected to acid deposition", *Philosophical Transactions of the Royal Society*, 305 (1984), pp. 353-368; D.L. GODBOLD, E. FRITZ, A. HÜTTERMANN: "Aluminum toxicity and forest decline", *Proceedings of the National Academy of Sciences (USA),* 85 (1988), pp. 3888-3892; D.L. GODBOLD, A. HÜTTERMANN (Eds.): *Effect of acidification on forest processes*, New York (John Wiley) 1994.

3 A. HÜTTERMANN/B. ULRICH (1984).

possible. So we used any outlet to the public which was at our disposal: TV, radio, lectures to any public where we could speak, including even gardener clubs.

We were highly criticised for this by the scientific community. On the other hand we were quite successful with regard to the impact of our research on the reduction of certain emissions. The most radical reduction based on our research findings was the lead of the petrol of cars. Leaded petrol has disappeared very fast and almost completely. Another success - story was the reduction of sulfur dioxide emissions. Fig. 1 illustrates the impact of our "campaign": the curve of the sulfur dioxide emissions on the territory of the former western part of the Federal Republic of Germany in the years since 1966. Until 1983, the curve shows a small overall decline of a total of 20 % in this time span. Since the GNP has increased during this period by 180 %, this indicates that the German Industry already did quite a lot to reduce its emissions. After that year, the curve bends down, ending at about 20 % of the value it had at the beginning of the eighties. This additional reduction was forced by the new regulations which were set up and enforced since 1984 in view of the results of "Waldsterben-research". Since some power plants still can be exempted from the regulations set up in 1984, this steep decline can be expected to continue during the next years. From some three millions tons of sulfur dioxide, the industry came down to 600 thousand tons.

In addition, in 1985 liming of the forests started at a rather high level, with a maximum of about 400 thousand hectares of forest that have been limed during annually at the peak of liming activities. This figure is now down at a constant level of about 300 thousand hectares of forest which are limed to stop acidification. So the research on "Waldsterben" lead to investments of more than one hundred billion Deutsche Mark for the desulfurication of the furnace exhausts, the costs for liming are now about hundred million Marks per year. So our research really set off economical changes which are quite considerable. This would not have been possible with our "march through the media".

II. Socio-Biology as a Threat Against Religion, Culture, and Society

Why is the message of this introduction relevant for the topic of this symposium? I strongly believe that there are times when you have to stand up and do much more in a right form than simply to publish papers and let it work. The topic of this symposium and the topic of at least a substantial part of the work our host has done during the last ten years is criticism of the concept of socio-biology. Socio-biology [4] in my opinion as it is put forward by Dawkins and all these people is threatening not only our religion but also our culture and our civilisation. One should not forget that the two totalitarian ideologies of this century, communism and fascism, were in their views of the world greatly influenced by the concepts of Darwinism of the last century. I would like to mention only the names of Haeckel and Bölsche in Germany [5], which influenced strongly both Engels and Lenin[6] and Hitler[7] in their basic ideologies.

Simply I have great objections against the ideology of Socio-biology and I found out that the most interesting and most relevant criticism against this concept comes from the organizer of this symposium and the group where he

---

4 The definition definition of sociobiology used in this communication means basically the ideology outlined in the following publications: E. O. WILSON: *Sociobiology. The New Synthesis*, Cambridge (Harvard University Press) 1975; R. DAWKINS: *The Selfish Gene*, Oxford (Oxford University Press) 1976 and their epigones.

5 The most influential publications being E. HAECKEL: *Die Welträtsel, gemeinverständliche Studie über monistiche Philosophie,* Leipzig (Kröner) 1899; W. BÖLSCHE: *Natürliche Schöpfungsgeschichte*, Berlin (Reimer) 1875.

6 C.f. F. ENGELS: *Herrn Eugen Dührings Umwälzung der Wissenschaft*, Leipzig (Genossenschafts-Buchdruckerei) 1878, W.I. LENIN: *Materialismus und Empiriokritizismus*, Moskau ("Sweno"-Verlag) 1909, for the impact of these classics on the socialistic society see e.g. R. HAVEMANN: "Die Einheitlichkeit von Natur und Gesellschaft", in: G. BUSCHENDORF, H. WOLFGRAMM, I. RADANDT (Eds.): *Weltall, Erde, Mensch*, Leipzig (Verlag Neues Leben) 1956 (this book was given to every person in the former DDR which at the age of forteen attended the "Jugendweihe", the socialist replacement of the religious ceremonies like Bar-Mitzva, Confirmation or First Communion)

7 See: Ä. BÄUMER: *NS-Biologie*, Stuttgart (Hirzel) 1990.

comes from[8]. The philosophically based criticism of this group introduces a completely different point of view into the discussion about socio-biology and shows the basic incorrectness in the whole concept.

The question I asked myself was whether this criticism gets its way through to the scientific community.

III. Methodology

The usual way to find out about the impact of any scientific publication is simply to consult the Science Citation Index. Unfortunately the publications listed in footnote 8 do not exist in this Index. This means the authors of the leading scientific journals do not quote, i.e. take notice, of the work mentioned there.

The discussion in science fortunately does not take place only in the papers covered by the SCI. There are in addition still highly influential journals where such discussions also may take place and which are not screened by the SCI, the two most important being *Scientific American* and *The Sciences* (the first is the oldest journal on popular sciences, the latter the journal of the New York Academy of Science, the biggest organisation of scientists in the world). I use in this communication therefore as indicators for any impact on the scientific discussion the letters to the editor concerning major recent articles on socio-biology published before in those journals.

For an evaluation of the data given below, it should be noted, however, that the usual reaction of the readers to the publications in these journals is rather low. A typical issue of *Scientific American* or its German translation *Spektrum der Wissenschaften* contains about eight main publications. Only few get any reaction in the "letters to the editor" - section, with one or two letters at best. So the general reaction of the public towards a publication in

8 The most important publication of the philosophically based criticism of Evolutionism and Sociobiology are: P. KOSLOWSKI, P. KREUZER, R. LÖW (Eds.): *Evolution und Freiheit*, Stuttgart (Hirzel) 1984; R. SPAEMANN, R. LÖW, P. KOSLOWSKI (Eds.) *Evolutionismus und Christentum,* Weinheim (Acta Humaniora, VCH) 1986; P. KOSLOWSKI: *Evolution und Gesellschaft*, Tübingen (Mohr Siebeck) 1984, 2nd edition 1989; R. LÖW (Ed.): *Bioethik*, Köln (Communio) 1990.

these journals is almost zero. Normally, the readers accept or forget what is written there, whatever they do - they do not write letters. That means on the other side, if there are several critical letters to the editor, this has a high significance. It reflects such a dissenting view of the reader that he behaves rather untypically and writes such letters of disagreement.

In this paper, the reaction of the scientific community is analysed by using four articles on four different socio-biological topics:

1. A field study concerning human behaviour in families and its interpretation
2. A preview of the book by Dennett: *Darwin's Dangerous Ideas*
3. A preview of the newest book by Dawkins: *River out of Eden*
4. A review of the meeting of the Human Behaviour and Evolution Society.

IV. Reactions on the Interpretations of a Field Study on Human Family Behaviour by Hemminger

The interpretations concerning human behaviour and its interpretation appeared in *Spektrum der Wissenschaften* in two articles, one by Hemminger and one by Voland. The papers deal with the same data base, an analysis of personal data of certain villages in Ostfriesland (Northern Germany) from the 17th to the 19th century, including 16.320 families with about 50.000 individuals, published in a book by Voland[9].

The first article by H. Hemminger[10] analyses the data of Voland and comes to the conclusion, that the socio-biological interpretation of those data is incorrect. Hemminger is a biologist, a PhD- student of Hassenstein in Freiburg im Breisgau and Privatdozent in Biology. For Hemminger sociobiology is a myth indeed.

9 E. VOLAND (Ed.) *Evolution und Anpassung*, Stuttgart (Hirzel) 1993.

10 H. HEMMINGER: "Soziobiologie des Menschen - Wissenschaft oder Ideologie" (Sociobiology of Human Beings - Science or Ideology), *Spektrum der Wissenschaften,* 6 (1994), pp. 70-80.

The reaction of the readers of this journal to his article was: nothing. No letter was published in subsequent editions.

One year later, Voland himself published an article in the same journal using the same data base that was analysed already one year before by Hemminger[11].

1. Basic Statements of Voland's Article

The statements are:

As any other feature of humans, the relationships between parents and their children are influenced by natural selection. When parents in this relation - perhaps unconsciously - apply economic criteria or egoistic interests, they act according to their evolutionary heritage.

In farmer families, the survival of male babies was dependent:

1. on the amount of available agricultural land (positive correlation)

2. the number of brothers already in the family (negative correlation)

The survival of female babies was negatively correlated to the number of sisters.

2. Reactions to Voland's Article[12]

Nine letters to the editor were published, all criticised Voland's article very severely as follows:

1. Out of the path of logical thinking (not reasons given why, however)

2. Bad and even falsifying statistics (in some figures, the ordinate does not start at zero)

3. Bad statistics, wrong conclusions, not only genes are inherited but also resources

4. Socio-biology does not give additional information for the interpretation of concrete examples and interactions.

5. Wrong data base, wrong scientific background, not enough data to draw conclusions

11 E. VOLAND: "Kalkül der Elternliebe - ein soziobiologischer Musterfall" (Calculation of parental Love - a perfect Example for Sociobiology), *Spektrum der Wissenschaften*, 6 (1995), pp. 70-77.

12 Cf. *Spektrum der Wissenschaften*, 4 (1996), pp. 6-9.

6. No consideration of psychological factors
7. Unjustified induction, theory not falsifiable.

Conclusions: it is urgently necessary to start a dialogue between philosophers and scientists. Important scientists mentioned: G. Roth, H. Schwegler, H. Maturana, F. Varela, St. J. Gould, L. Margulis, J. Lovelock, G. Bateson, W. Schwemmler, F. Wallner, M. Weingarten, K. Edlinger, W. F. Gutmann.

8. Wrong treatment of data, not all possible aspects of human infant survival are taken into consideration. The conclusions are of the same quality as the following statement: "The study has shown that in the present territory of the United States the ownership of guns and a white skin of their owners have a positive influence on the rate of reproduction".

9. Scholars of the Social Science would interpret the data completely differently and the socio-biological interpretation of the data is not stringent.

3. Reply by the Author Voland: none

(in this journal, an author gets the possibility to reply to critics)

V. Reactions to Previews of New Books in Socio-Biology by Dennett and Dawkins

We analysed the reactions of readers to the previews of two new books on socio-biology, Dennett's article "Darwin's dangerous idea."[13] and Dawkins's preview of his book *River Out of Eden.*

1. Basic Statements of Dennett's Article and the Reactions to it

a) Basic Statements of Dennett's Articles

Like "universal acid", evolutionary theory eats through every other explanation for life, mind, and culture.

13 D.C. DENNETT: "Darwin's dangerous idea", *The Sciences*, 35/3 (1995), pp. 34-40; German Translation: "Es geht auch ohne Gott", *DIE ZEIT,* Nr. 8, 23. Februar 1996, S. 33.

Evolution "works" via algorithms. Three key features are of importance here: substrate neutrality, underlying mindlessness, and guaranteed results. No design is underlying evolution.

God is, like Santa Claus, a myth of childhood, not anything a sane, undeluded adult could literally believe in.

b) Reactions to Dennett's Article in *The Sciences*

The reactions to the article in *The Sciences*[14] consisted in nine letters: one neutral (about the "history" of universal acid), the others disagreeing with his view.

Basic criticism:

1. Annealing is no correct model for evolution and no algorithm. Evolution is far from being understood.

2. Complete disregard for the history of biology. The algorithm analogy shows the author being out of touch with modern scientific thought.

3. The accidental occurrence of the basic biochemical system of living beings is not reasonably possible within the mass of the known universe nor its time span.

4. 5. Algorithms are not a fool-proof recipe. Values cannot be based on a new scientific viewpoint alone. Dennett is unaware of the modern debate between science and religion (Kant).

6.7. Algorithms are an inadequate model. Nature is too complex to be reduced to such a simple model.

8. Who has made the laws of nature ?

c) Reply by Dennett

Basically, he replied: "Read the book!"

d) Reactions to Dennett's Article in *DIE ZEIT*

The reactions to the article in *DIE ZEIT* consisted in eight letters: one positive, the others disagreeing.

Basic criticism:

1.2. Dennett's theory cannot explain "last things": who was responsible for the big bang, etc. "Eternal life" cannot be falsified by Darwinism.

14 Cf. *The Sciences*, 35/5 (1995), pp. 6-7, 48.

3. The level of argument of Dennett has the same quality as a chemist's "analysis" of a poem: It is nothing more than paper and ink. His statement is not falsifyable, but does not explain the whole story.

4 - 6: Who defined the laws of nature, who defined the algorithm.

7. Simply bad philosophy at a low intellectual level.

2. Basic Statements of Dawkins's Article and the Reactions to it

The third article is a preview by Dawkins of his book *River out of Eden*[15].

a) Basic Statements of Dawkins's Article

Life has no higher purpose than to perpetuate the survival of DNA.

Nothing can stop the spread of DNA that has no beneficial effect other than making males beautiful to females.

As long as DNA is passed on, it does not matter who or what gets hurt in the process. Genes don't care about suffering, because they don't care about anything.

DNA neither cares nor knows. DNA just is. And we dance to its music.

b) Reactions to Dawkins's Article

The reactions to the article consisted in four letters[16], all disagreeing:

1. Full of errors of logic. Substituting "utility function" for "purpose" makes the argument not more valid.

2. Dawkins is hijacked by his own metaphors. While denying supernatural design, he is teetering on the brink of attributing some pretty malevolent characteristics to nature.

3. Dawkins's argument "the purpose of DNA is to perpetuate life and thus itself" is circular and unsatisfying like "the purpose of a V8-engine is to make money for General Motors".

4. DNA explains much but not everything. It is evident that matters of ethics and aesthetics in human society do not dance to the music of the double helix.

---

15 R. DAWKINS: "Gods Utility Function", *Scientific American*, 11 (1995), pp. 62- 67. Preview to his book *River Out of Eden*.

16 Cf. *Scientific American*, 3 (1996), p. 5.

3. Reactions to a Review of a Conference of Social-Darwinists

The fourth article analysed in the response it received from its readers is a review on the conference of the Human Behaviour and Evolution Society held at Santa Barbara which was prepared by a staff member of *Scientific American*.[17]

Basic statements of the article:

Natural selection, the decisive principle in Charles Darwin's Theory of Evolution gets more and more accepted in anthropological and psychological disciplines, which understand spiritual achievements, speech, and differences in sex as biological adaptations.

Reactions to the article:[18] five letters, all disagreeing.

1. The Darwinistic concept is not able to describe humans in a complete way. All those attempts must necessary fail in view of the complexity of the human nature and the always existing possibility of self-determination.

2. Some of the basic features of Social-Darwinism have long been falsified (like the unspecific aggressiveness claimed by Lorenz). The theory of genetic programmed human behaviour is an error.

3. Social-Darwinism ignores the power of culture.

4. Even if one accepts that evolutionary optimisation plays an important role in the development of human societies, one can doubt that the genes are the very basis for that. Culture probably plays a more important role. Even the very concept of the gene is more complex that the social-darwinists assume.

5. The way of dealing with evolutionary theories is trivial, its possible results may be horrible (fascist and communist Social-Darwinism)

VI. Conclusions

The analysis of the response of readers of popular scientific journals both in the USA and in Germany is very interesting:

17 Cf. *Spektrum der Wissenschaften*, 12 (1995), pp. 80-88.

18 Cf. *Spektrum der Wissenschaften* (1996), pp. 8-11.

1. Papers using the socio-biological ideology as a basis for their research and their conclusions are under relatively hard "attack" from disagreeing readers, whereas in the example of a full paper criticising this ideology, no reaction took place.

2. The readers writing critical letters to the editors are not aware of the philosophically based criticism put forward by Löw, Spaemann, Koslowski and others (see footnote 7).

There is obviously a great concern in the scientific community about socio-biology, otherwise the papers on socio-biological issues would not have been received so controversially. On the other hand, people that have bad feelings about this threatening ideology do not have the right arguments to deal with it.

The reason for this is, in my opinion, that the philosophically based criticism of socio-biology is not published in a way that reaches the scientific community concerned, i.e. it is not properly provided to them.

Figure 1. Relative emissions of sulfur dioxide and liming of forests in the aerea of the FRG before unification (data from *Umweltbundesamt* and Forest reports). The relative unit of sulfur dioxide emissions equals 3 million tons per year, the one of liming 400.000 ha of forests.

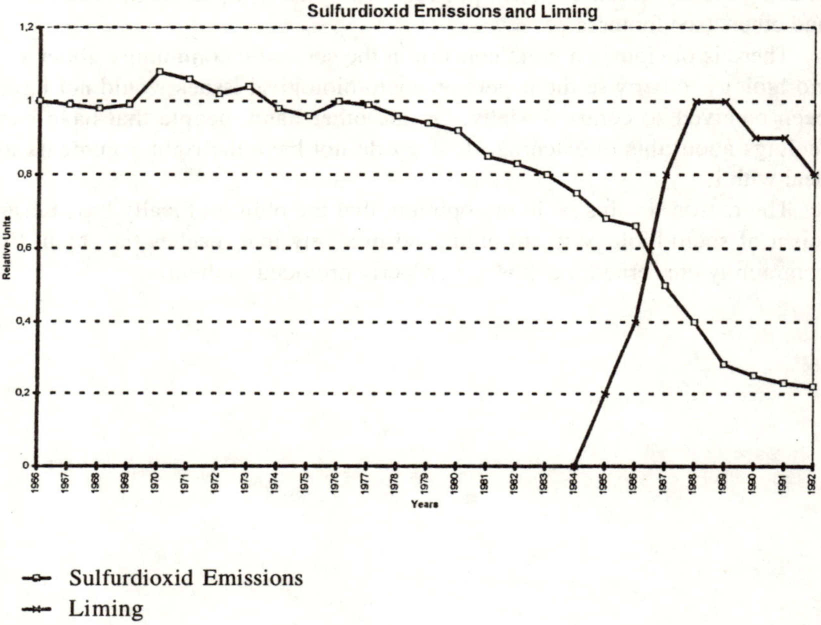

Part IV

Synergetics, Symbiosis and Cooperation in Sociology and Biology

Chapter 9

Synergetics in Sociology and Biology

HERMANN HAKEN

I. The Individual and the System

Each human being is embedded in some society. Here we take the word "society" in a wide sense and may understand by it a group of humans ranging from a family to the world population. In the following we want to study the relationship between the individual and the system from the point of view of synergetics, an approach that we shall outline below. When we consider an individual, there are two main factors that may determine his or her development. The one factor is the gene, where one extreme attitude is taken by Dawkins (Dawkins 1976) with his concept of the selfish gene. The other factor is the environment that acts on the individual in a variety of ways that we shall discuss below. Rather clearly, there is an interplay between the genetic outfit and the impact of the environment. Society being composed of individuals is by no means fixed, it behaves rather in a fluid manner. Moreover, a society does not only influence the individual, but the latter may also influence the society in a variety of ways. This leads us to the question of how we can characterize human social organization and what are the mechanisms that bring about such kind of organization. In my article I will adopt the attitude that society as well as any organism in nature underlies the laws of self-organization that have been unearthed by synergetics (Haken 1983, 1984). When dealing with complex systems, such as society, it will be a crucial task to characterize their collective features. Some manifestations of collective human behavior are certainly language and culture, but we would like to find means how to identify or characterize such collective features.

II. What Is Synergetics About?

In my article I wish to apply the methods and results of synergetics to biology and sociology. So I should first explain what synergetics is about. It is an interdisciplinary field of research that deals with complex systems that are composed of many individual parts. By their mutual interaction these parts can produce macroscopic structures, i.e. they can produce specific collective features. It may be surprising to the reader that the basic concepts underlying synergetics apply to a great variety of different systems, irrespective of the kind of the individual parts. They apply to systems in physics, such as fluids, to chemical reactions that form macroscopic patterns, to cells in a body, or to neurons in a brain, to mention a few examples. If thus the results of synergetics are applied to human behavior, this seems to be shocking to some readers and I will have to discuss this point in more detail in the course of this article. Because of its generality, the results of synergetics can be applied to all levels of biological organization, starting from, say, an individual cell to cell assemblies, or to all organisms. Quite clearly, there must be a price to be paid for the generality. The price is three-fold:

1. We describe the systems at a rather abstract level.
2. An adequate form of description is a mathematical theory.
3. Above all, we consider situations of systems in which there are qualitative changes on macroscopic scales. What is meant by "qualitative changes" and by "macroscopic scales" will become more obvious below.

Because of the study of qualitative changes, synergetics may be considered as a theory of the emergence of new qualitites. So in particular, we will see which new qualities can be developed by society. In spite of the abstract level, the results are quite astonishing and lead to, what I believe, new insights in the study of biological and sociological systems.

III. Basic Concepts of Synergetics

In order to make the results of synergetics understandable to a broad audience, I will refrain from any mathematical formulation, but rather attempt to describe these results qualitatively. One central concept is that of the *order parameter.*

Let us consider a system composed of individual parts and study the behavior of them. Then in a number of cases it may appear that these individual parts show common behavior. Examples are a group of people of which each member speaks the same language or each of which has the same political opinion. The order parameter is a quantity that acts like a puppeteer on the individuals. Or, in other words, the behavior of the individuals can be described by means of that specific order parameter. Instead of dealing with a great number of individuals, say, in a nation, it is sufficient to deal with the behavior of the order parameters. The action of the puppeteer that determines the behavior of the individual parts, is called the *slaving principle of synergetics.* I am aware of the fact that the word *slaving* is disliked by a number of sociologists who claim that humans can't be enslaved, they have a free will. I will come back to this delicate issue below. Instead of "slaving principle", one may also speak of *consensualization* though this word is, maybe, less indicative than the word *slaving principle.* There is, however, a basic difference between the results of synergetics on the relationship between order parameters and enslaved individuals and the puppeteer and his or her puppets. In contrast to the latter example, the individuals react on the order parameter and thus determine its behavior. In this way, the order parameter determines the behavior of the individuals, but in turn the individuals determine the behavior of the order parameter. We are speaking here of *circular causality.* Instead of describing the behavior of each individual, in this approach it is sufficient to deal with the order parameters, which actually can be represented by mathematical quantities that obey specific order parameter equations. According to these equations, order parameters can compete with each other, whereby only one order parameter wins the competition and determines the behavior of the whole system. Order parameters can, as we shall see, coexist, or they can even stabilize each other by cooperation or, as a biologist would say, by symbiosis. The amount to which order parameters act may underly a temporal change, for instance in the simplest case the action of the order parameter does not change over time, or order parameters may oscillate in their activity, or when there are several order parameters interacting with each other an entirely irregular temporal behavior may evolve that is nowadays called *deterministic chaos*. How can we distinguish between order parameters and the individuals in an abstract way?

As it turns out, order parameters change slowly over time, whereas individuals may change their behavior rapidly, or individuals may be born and die while order parameters exist. This distinction is called *time-scale separation*. Admittedly these notions are rather abstract and, therefore, time has come to explicate them by means of an example.

IV. Language as an Example for Order Parameters

In the relationship between language and individuals the time scale separation between order parameters and enslaved parts of a system is quite obvious. The language of a nation is much longer existent than any individual. When a baby is born, he or she learns the language and is - in terms of synergetics - enslaved by the language. When grown up, the adult carries on the language. Without individuals there would be no existing language. On the other hand, language determines the behavior of individuals. Language may fulfill a variety of tasks: it may serve for communication, but it also serves for the discrimination of different groups, or it serves for the identification of the members of a group. It is a remarkable fact that fine accents can be recognized and one may, for instance, say which part of New York a person comes from. The circular causality between language and individual is quite obvious. Quite in the sense of order parameters, different languages can compete with each other. An example is provided by the competition between English and German in the United States, where the English language won. They may coexist, for instance, in countries, such as Switzerland, where the coexistence is made in particular possible by the existence of different regions, where these languages are spoken. Languages may cooperate, for instance a dialect and a high language can help each other to survive.

Another important point, quite in the sense of synergetics, is the possibility of learning a language. It is well established that any natural language can be learnt by any individual stemming from another nation. Thus the genetic outfit does not predetermine the language to be learnt, the genetic outfit is sufficient so that any individual can learn any language. In a way the genetic outfit with respect to language acquisition can be called *totipotent*. The efficient action of language as an order parameter is underlined by experiments in the Middle Ages done by a Staufer king, who let children grow up without communication with other persons. These children never learnt the language and in addition they never were able to produce a language. Here we observe another property related to circular causality that may be called also the *egg-hen problem*. What is first, hen or egg? Or in our case: What is first, the existence of a language or the use of a language by an individual?

It is interesting to know that some kind of artificial language, such as Esperanto, can also be learnt, but it turns out that these languages have never come to real life. This is again connected with the order parameter concept. In order to produce the specific order parameter, there must be enough individuals who carry it and there must be enough opportunity to use language within a group. At this moment we can introduce another notion of synergetics, namely *symmetry* and *symmetry breaking*. The genetic outfit can be said to be symmetric with respect to any language. But when a specific language is learnt, this symmetry is broken. We shall come back to the phenomenon of broken symmetry at a variety of occasions below. We shall see that symmetry breaking is an important property of any system including biological and, in particular, sociological systems. This will have important insights into ethics, moral, and so on.

V. More Examples for Order Parameters

The kind of a state, such as democracy or dictatorship, determines the order parameter. The state is brought about by the collective action of its citizens, but, on the other hand, the state then determines the political behavior of its citizens. Hereby it is obvious that the behavior may be quite different, namely for instance in a democracy all sorts of political opinions are permitted and can be expressed, whereas in the dictatorship this is, of course, not the case. The public opinion coincides with that of the ruling class. However, it must be said that there is - even in the democratic case - an order parameter, namely the consensus to maintain democracy. Another example for order parameters is provided by rituals, where in the first step humans possibly subdue to the ritual with their free will, but later on obey it further and carry it on. This is, of course, still more pronounced when children are born in families that obey specific rituals. A further example is provided by culture or, say, styles of art, and so on, where the slaving principle works more or less efficiently and where we may quite often observe the coexistence of several order parameters, though in a number of cases we also find the effect of competition and one winning order parameter. After a little thought, we recognize that also tabus play the role of an order parameter. We are presently witnessing the occurrence of a new order parameter in the United States, namely that of political correctness. When we take the order parameter concept and the slaving principle connected with it serious, we are also led to consider ethics and religion as order parameters. In a way, this may even seem obvious. But there are far reaching and disturbing further conclusions, when we take the concept of order parameters and of their properties serious. Namely as we have seen by means of the example of languages, order parameters are by no means uniquely determined. They rest, in a way, on collective consensualization on issues that are a priori by no means uniquely fixed. Our theme is by no means only abstract, but is witnessed by the collective suicide of some religious groups, or of fanatic religious behavior, where the world abounds of. The same is true for ethics. Just look at the issue of the death penalty to mention a visible example, though there is a great variety of far more subtle issues, in which no unique solution has been reached. Quite generally, it must be stated that in self-organizing systems there are several order parameters possible and below we have to study which causes act in order to fix a specific order parameter by symmetry breaking. There we shall have to study the question, how order parameters or, in other words, their collective effects are brought about. Connected with this is the question whether there are definite reasons that determine one specific order parameter. Let us consider ethics and religion specifically. When we look at these issues from the abstract level of synergetics, we can find only few reasons for the privilege of one specific order parameter:

1. The symmetry is already broken by each individual even if isolated from society. This would imply that in the case of humans their genetic outfit provides them with a specific order parameter.

2. By some kind of collective selection a group, for instance with a specific ethics, is more successful than another group with a different ethics. This point of view was promoted by the famous economist Friedrich August von Hayek, (von Hayek 1949) who stated that such ethics has evolved in which the citizens could develop a more efficient economy.

3. The third possibility that is also taken care of by synergetics, is that symmetry is broken from the outside by the interference of an outer force. In the case of religion, we are led then to divine revelations and it is, perhaps, not surprising that divine actions are crucial ingredients of great religions.

I am fully aware of the fact that the whole question is a delicate issue, but if we start to look at these problems from an abstract point of view, they naturally pop up.

VI. Communication and Synergetic Interrelational Networks

Is the existence of order parameters, as outlined in the foregoing sections 4 and 5, just the outcome of mathematical relationships, or are there more visible mechanisms behind their formation? In this section I wish to study some of the mechanisms.

First of all, the individual parts of a system must interact with each other, which would mean for the human society that its members must communicate with each other. Such a communication may be at the level of a person within small groups, or via mass media. Let us consider the ways in which these interactions may take place. For me a crucial experiment was done by Solomon Asch (Asch 1952) in the thirties, an experiment quite well known to sociologists. Asch had a number of subjects, say ten, who were sitting in a row, but actually only one subject was a true test person, while the other nine subjects were his helpers. Asch then showed three lines of different lengths to these persons and they had to tell which line was the longest. Then in the first test run all the helpers gave the correct answer and so did the test person. In a second run his helpers "identified" a wrong line as the longest and in about 60 per cent of cases the true test person changed his or her opinion accordingly. For any mathematical modeling in synergetics this experiment is crucial, because it shows that other persons can influence the member of that group. It must be said, however, that the interpretation of this experiment by sociologists is not unique. Some of them say that the test person just wanted to be sociable and thus agreed to the publicly uttered opinion. These sociologists say: If the issue had been more important, this person would have stuck to his or her original opinion. I must say that I do not share this interpretation, because in difficult social or political questions the situation is so highly complex that it is hard for an individual to take a unique position; rather individuals are then inclined to follow the opinion of the majority, or, at least, of a larger group. Quite clearly, this issue needs further clarification, but in the very moment we admit that individuals can be influenced by others to some extent, the existence of order parameters can be shown mathematically. Another communication means is, of course, mass media. They seem to be predestined as order parameters, and then, in the parlance of synergetics, to enslave the members of the society. But in discussing this point, we must not overlook an important property of order parameters and the slaving principle, namely that of circular causality. There is not only the action of the order parameters towards the individuals, but the individuals generate the order parameters by their joint action. This circular causality principle can nicely be seen in the mass media at least when they depend on consumers. When, for instance, a newspaper is bought, then there is already some willingness of the consumer to buy this special one and we may presume that this is only the case if the newpaper fits more or less the opinion of the consumer. This is, of course, too superficial a statement, but it helps to illustrate my point, namely that the collective behavior of the consumers determine the survival or perish of a specific newspaper.

The issue is more complicated with government-owned television stations, where there is no feedback of the consumers, maybe with exception of some control agency, which reflects, for instance, the percentage of the political parties, trade unions, etc. Further important issues in this general context are the existence of mental maps, or, in other words, to be more specific, how an individual perceives his or her environment. I just mention as an interesting example field experiments by Portugali done in Israel, where he asked Arabs or Jews about the nearest settlement. (Portugali 1993). As it turned out, the Arabs mentioned another Arab settlement as the nearest in spite of the fact that a Jewish settlement was closer and vice versa. Portugali also established the notion of interrelational networks, (Portugali, in press) a question that was studied further by Portugali and the present author. (Haken/ Portugali, in press) He emphasizes that humans do not only store knowledge internally in their memory, but that they are more and more utilizing external means of storage, and more recently, of processing information by computer networks, such as the world-wide web. As our studies show, here new order parameters evolve and structure the information landscape. It will be beyond the scope of this article to go into more details.

VII. Pattern Recognition and Decision Making

As transpires from the previous section, in establishing social connections the question how humans perceive their surrounding and how they make decisions plays a fundamental role. Again lack of space does not allow me to enter these problems in any greater detail, but I just wish to report on some of our experiments by means of the synergetic computer for pattern recognition. (Haken 1991) These computer experiments, aimed at mimicking pattern recognition by humans, indicate that attention plays a crucial role. If we are not prepared by attention to recognize a specific pattern, we are unable to do so. Attention is strongly influenced by emotions, such as hopes or fears, hormons, and so on. Pattern recognition can depend on previous experience as is witnessed by Fig. 1, which illustrates the so-called hysteresis effect. Another effect that illustrates the oscillations of order parameters is provided by ambiguous figures, such as Fig. 2. Here one may first recognize the vase in the middle. This percept then fades away and gives way to two faces at the left and right sides. Then their perception fades away and we again see the face, and so on. In this case one may ascribe one order parameter to the vase and another one to the two faces, and quite evidently these order parameters oscillate. I should like to promote the thesis that pattern recognition gives us some clues about *cognitive processes*. As an example I take here the question of how one makes decisions. The hysteresis effect means: one makes the same decision as before in spite of the fact that circumstances have changed. Oscillation means: we are oscillating back and forth between two options. These analogies that are sketched here rather superficially can be carried further. (Haken 1995b)

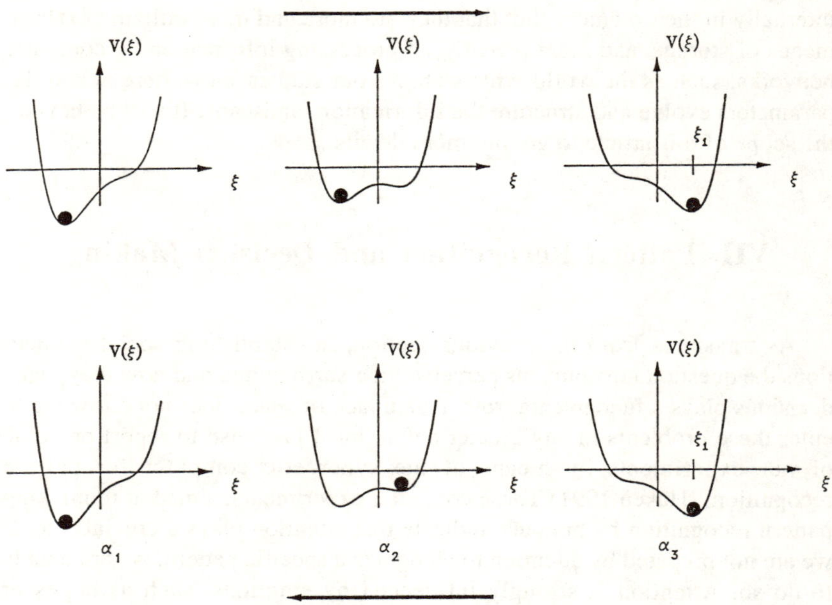

Fig. 1: The hysteresis effect. Upper part, first row: The state of a system is described by the order parameter, whose dynamics is visualized by means of the movement of a ball in a hilly landscape. According to the mathematics ofsynergetics, a hilly landscape can be deformed by means of the change of a suitable control parameter. We indicate three steps of different forms of a landscape when a control parameter is changed from left to right. In the middle the ball is kept at the same position as before, but in the right picture the ball had fallen to another valley. This may be considered as a metaphor that the macroscopic state of the system has changed. When we now look at the second row, going from right to left, we see that the ball now will stay in the

same position as in the right picture, and only in the left part of the picture it will acquire its previous position. The essential issue of this very simple visualization is this: In spite of the fact that the landscape for the order parameter is the same in the middle picture, the state of the system characterized by its position in one or the other valley is different depending on the history. This is the effect of hysteresis. The lower part of this picture shows that even our brain shows this effect. Look at the upper row, starting from the left with the face of a man, follow this line to the right, then start again from the left and follow up the line to the right. Then you will see a switch from the perception of a face of a man to a kneeling woman. When you reverse the whole procedure, the switch from the perception of a kneeling woman to a man's face will occur in the upper row.

Fig. 2: An example for an ambiguous figure, where the perception vase/faces oscillates back and forth.

VIII. What Causes Changes

In the preceding sections I considered order parameters as given. On the other hand, quite clearly, in the course of time such order parameters must have come into existence in some way or another. In the following I wish to describe the approach of synergetics to this quite fundamental problem. The basic idea is that systems are embedded in environments.

Examples of such environmental conditions may be the temperature on earth, or the influx of energy to the earth from the sun, the amount of food that is available to a population of humans or animals, and so on. In biological systems we have also to deal with internal environments; so from a certain point of view the level of hormons or neurotransmitters in a body or brain, respectively, provides such an environment for an organism. These environmental conditions are characterized or even quantified by *control parameters* in synergetics. The general point of view adopted by the mathematical theory of synergetics is as follows: We start from a given set of control parameters at fixed values. Under these conditions a certain state of the system under consideration will have been established. This may be a stationary state, i.e. one that does not change over time, but it may be also an oscillatory state, and so on. When we now change one or several control parameters, at least in general, the systen does not qualitatively change its state. In technical terms we are speaking of structural stability. Structural stability can be visualized by means of an example taken from biology (cf. Fig. 3). (Thompson 1961) When we change one or several control parameters more and more, in a great variety of systems quite surprisingly a new phenomenon occurs, namely a sudden, qualitative change (Cf. fig. 4).

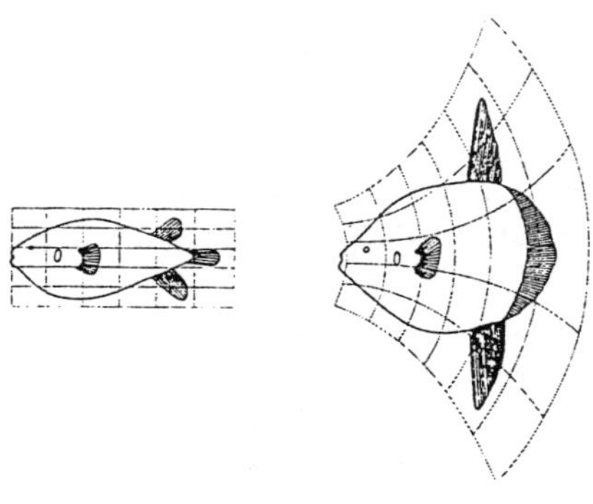

Fig. 3: Visualization of structural stability as d'Arcy Wentworth Thompson had shown. These two kinds of fish can be obtained by a mere deformation of a coordinate system, where no qualitative changes occur.

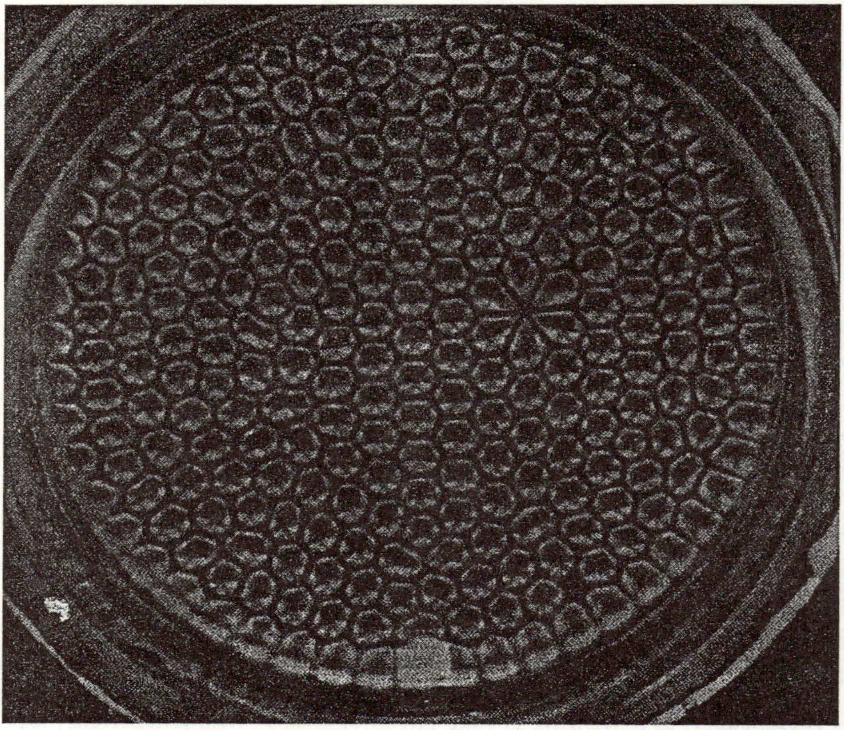

Fig. 4: Example of a qualitative change. When a liquid in a circular vessel is not heated, it is at rest and shows a homogeneous state. When the liquid is heated from below and cooled from above, under specific conditions a honeycomb pattern spontaneously develops. The system has changed its macroscopic state qualitatively.

Quite a different structure is generated (Fig. 4). As has been unearthed by synergetics, it is there where new order parameters appear and substitute the former ones.

Because of circular causality, the occurrence of new order parameters is by no means obvious: the establishment of order parameters requires the cooperation of the individual parts of the system, and then only the order parameters can react on the individual parts by means of the slaving principle. By means of a number of examples that originally were taken from physics, for instance laser physics, but then could also be substantiated by the analysis of electric and magnetic brain activity, the mechanism is as follows: When a control parameter surpasses a critical value, in the first instance the system becomes destabilized and seeks a new stable state. Close to its transition point there are large fluctuations in the behavior of the individual parts of the system. These fluctuations indicate that the system tests the possibility of establishing a variety of different order parameters. This phenomenon is well known in physics and is called *critical fluctuation.* In this region any perturbation of the system drives it away from its equilibrium state and it takes the system a considerable time to relax towards its former equilibrium state - a phenomenon called *critical slowing down.* The important phenomena of critical slowing down and critical fluctuations can be well observed at the scales of sociological and economic processes as is witnessed by the processes accompaning the decay of the former Soviet Union. When the control parameters are further, say, increased, the system may reach a new stable point. However, and this is quite crucial, as synergetics shows, in the general case beyond a critical point there are several possibilities for the order parameters to adopt. Which one of the order parameter values is then adopted by the system, depends on chance events that cannot be predicted by theory. Thus a small fluctuation decides a later macroscopic outcome, i.e. the establishment of a definite order parameter that enslaves the whole system. A couple of years ago I suggested this mechanism as one valid for revolutions, a point of view that had been anticipated, as I learnt later, by Lenin. According to this point of view, in order to generate a revolution one has first to destabilize a system, for instance by terrorism, and then in the destabilized situation a decided group of people must take the lead and drive the system into a new stable state. The chance event here is, if course, that, in general, one cannot predict the behavior of individuals in extreme situations. The whole approach adopted here sheds new light on the causation of social processes. It is not the direct command that drives humans into specific actions, but indirect controls set by, say, economic necessity, and so on.

HERMANN HAKEN

IX. The Selfish Gene Revisited and the Theory of Evolution

The concepts of synergetics allow us to shed new light on Dawkins' notion of the selfish gene. (Dawkins 1976) Let us consider the time scale criterium that allows us to distinguish between order parameters and the enslaved subsystems. Quite clearly, the gene, or rather the gene pool, is much longer living than any individual. This suggests to interpret the gene pool as order parameter, that, indeed, enslaves the behavior of each individual. However, and there is another aspect to the whole issue, namely here circular causality comes in again. The *order parameter*, in the present case the gene pool, is determined by the cooperation of the individual parts, this time mainly not in space but in time. Thus if the gene pool does not sufficiently equip the phenotype (the individual) for survival, the whole gene pool will die out. Interpreting gene pools as order parameters, we immediately can apply the concepts of competition, cooperation and coexistence, though these phenomena are termed differently in biology, namely natural selection, symbiosis, and ecological niches, respectively. This brings me to the issue of the theory of evolution. Though we know that there are a number of variants, this theory is essentially based on the action of two processes, namely mutations and natural selection. Here the identification with the concepts of synergetics can be made as follows: The number of the individuals of a species corresponds to an order parameter, mutations correspond to fluctuations, i.e. to the spontaneous formation of order parameters, and natural selection corresponds to competition between order parameters as was just outlined. In this respect it is worth mentioning that Eigen's theory (Eigen 1992) of the evolution of biomolecules from a primordial soup is the most advanced mathematical formulation of a theory of evolution. Surprisingly there is a one-to-one correspondence between Eigens' equations and those of some physical systems, such as lasers. (Haken 1981) Such mathematical theories allow one to cut the Gordian knot that consists in saying that the "survival of the fittest" is a tautology. In these mathematical formulations one can, at least in principle, calculate what is called *fitness* by Eigen and thus predetermine which kind of species will win the competition. I am fully aware of the fact that my brief remarks here are rather superficial, but they may indicate that by means of the general concepts of synergetics we may establish far reaching analogies and thus draw conclusions even on evolutionary processes. There is one basic difference, however, between Eigen's theory of evolution and the ideas of synergetics. Namely in Eigen's theory evolution goes on continuously, i.e. without specific interference by the environment. In the notion of synergetics, the environment interacts crucially with the evolution of new species because of the distinguished role of control parameters, whose change may trigger an instability of a system that in the present case would mean that suddenly new species evolve. Thus in this synergetic view, evolution is driven again and again by changes in the environment, in particular climatic changes. It would be worth to discuss within this frame the notion of punctuated equilibria as introduced by Eldredge, but it will be beyond the scope of this article to go into more details.

X. Dynamics of Settlements

Let us make a big jump from the question of genes and evolution to a macroscopic phenomenon of social interactions, namely the dynamics of settlements. I refer here, in particular, to Weidlich's analysis (Weidlich/Haag 1983) of the cooperation and competition of social groups within a city. Let us consider the following situation: In a city there are two kinds of populations, A and B. In order to study the way they try to live together or to separate, Weidlich introduced a number of parameters that in our interpretation can be called control parameters. One such parameter is the affinity within group A or B, respectively, which means that there is a tendency of the members of each group to live together. There is another parameter, γ, that tells the affinity of group A to group B, which may be also repelling, and of group B to group A, called δ. For a certain set of these parameter values, it turns out that both groups mix, i.e. they live in the same part of a city. When the parameter γ is changed slightly, these groups separate from each other. When γ and δ are changed beyond critical points, in this model it will happen that these groups are constantly moving from one quarter to another, where group B tries to penetrate group A, group A leaves that quarter, group B settles down there, and so on. If we would imagine a city as a circle, a circular movement would occur here. Such phenomena are, of course, observed in a number of cities, especially in the United States. But what is interesting in this model is the following issue: The way in which the groups A and B behave, critically depends on the values of the individual interaction parameters. Thus slight changes in human relationships or general attitudes may cause dramatic qualitative changes of the whole behavioral pattern. On the other hand, and this is the disturbing issue here, it is, in general, very hard or even impossible to measure these parameters. This, at the same time, shows the advantages and limitations of these kinds of models. While these models give us new insight into the qualitative behavior of social groups, it is impossible or rather difficult to make precise predictions because of the lack of knowledge of the parameter values and their possible changes.

XI. Cooperation and Coherence of Living Organization

The title of this section is taken from the title of a contribution to these proceedings by Mae-Wan Ho (Ho, these proceedings). Unfortunately, I haven't had the opportunity to familiarize myself with his work. I apologize for my lack of knowledge, but at the same time I hope that we can start a discussion on differences and common features of our approaches. Coming close to the end of my article, I want to make a few general remarks on the application of the principles of synergetics to biological systems. As it may have transpired from what I have outlined above, the principles of synergetics apply to a wide range of biological and sociological phenomena. Having in mind the time scale separation between different processes in a complex organism, we may readily state that the principles of synergetics apply to all sorts of levels in the biological hierarchy, from the metabolism within a cell to the cooperation of many cells in the brain. Indeed, more recent analyses of ours (Fuchs/Kelso/Haken 1992, pp. 917-939; Jirsa/Friedrich/Haken/Kelso 1994, pp. 27-35; Jirsa/Friedrich/Haken, unpublished; Jirsa/Haken, Submitted To Physica D) of MEG Experiments (Kelso/Bressler/Buchanan/Deguzman/-Ding/Fuchs/Holroyd 1991) and EEG Experiments (Fuchs/Friedrich/Haken/Lehmann 1987; Friedrich/Uhl 1992) have shown that in the brain there are pronounced qualitative changes of macroscopic patterns of activities of its neurons, where their different collective states are again and again governed by specific order parameters, whose dynamics can be analysed in a number of cases. It appears more and more that the brain operates close to instability points. This allows it to adapt quickly to new situations. A number of behavioral patterns become explicable in this manner. It will be beyond the scope of this article to go into any more details, but I wish to draw the readers' attention to a recent book of mine "Principles of Brain Functioning" (Haken 1996). Along these lines, we may conceive the whole biosphere or even the whole global system as a giant synergetic system that obeys the principles of synergetics. In other words, we may and must expect that at critical points of control parameter values small changes of parameters can cause dramatic changes in the overall behavior of, say, the biosphere. This sheds again some light on the issue of climatic catastrophes, for instance possibly caused by the $C0_2$ level, an issue that would require another article, however.

XII. Concluding Remarks

In my article I have tried to draw a coherent picture of how synergetics allows us to look at quite different phenomena in biology and sociology from a unifying point of view. The most remarkable result of our analysis is the fact that systems may undergo dramatic macroscopic changes at certain critical control parameter values, which should make us cautious in dealing with our environment and also with our human relationships.

Bibliography

ASCH, SOLOMON, E.: "Group Forces in the Modification and Distortion of Judgements", *Social Psychology*, 452, New York (Prentice Hall Inc.) 1952, pp. 112-117.

DAWKINS, R.: *The Selfish Gene*, Oxford (Oxford University Press) 1976.

EIGEN, M.: *Stufen zum Leben*, Oxford (Oxford University Press) 1992.

FRIEDRICH, R., UHL, C.: "Synergetic analysis of human electroencephalograms: Petit-mal epilepsy." in: R. FRIEDRICH, A. WUNDERLIN (Eds.): *Evolution of dynamical structures in complex systems*, Berlin (Springer) 1992.

FUCHS, A., KELSO, J.A.S., HAKEN, H.: "Phase transitions in the human brain: Spatial mode dynamics", *International Journal of Bifurcation and Chaos* , 2 (1992) pp. 917-939.

FUCHS, A., FRIEDRICH, R., HAKEN, H., LEHMANN, D.: "Spatio-temporal analysis of multichannel α-EEG map series", in: H. HAKEN (Ed.): *Computational systems - natural and artificial*, Berlin (Springer) 1987.

HAKEN, H.: *Light, vol. II, Laser light dynamics*, Amsterdam, New York (North-Holland-Elsevier) 1981.

HAKEN, H.: *Synergetics, An Introduction*, 3rd. ed., Berlin (Springer) 1983.

HAKEN, H.: *The Science of Structure*, New York (Van Nostrand-Reinhold) 1984.

HAKEN, H.: *Synergetic Computers and Cognition*, Berlin (Springer) 1991.

HAKEN, H.: "An Application of Synergetics. Decision Making as Pattern Recognition", *Zeitschrift für Wissenschaftsforschung*, Graz (1995).

HAKEN, H.: *Principles of Brain Functioning, A synergetic approach to brain activity, behavior and cognition*, Berlin (Springer) 1996.

HAKEN, H., PORTUGALI, J.: "Synergetics, Inter-representation networks and cognitive maps", in: J. PORTUGALI (Ed.): *The Construction of Cognitive Maps*, Dordrecht (Kluwer Academic Publishers) in press.

VON HAYEK, F.A.: *Individualism and Economic Order*, London (Routledge and Kegan Paul) 1949.

HO, MAE-WAN: these proceedings.

JIRSA, V.K., FRIEDRICH, R., HAKEN, H., KELSO, J.A.S.: "A theoretical model of phase transitions in the human brain", *Biol. Cybern.*, 71 (1994), pp. 27-35.

JIRSA, V.K., FRIEDRICH, R./HAKEN, H.: *Reconstruction of the spatio-temporal dynamics of a human magnetoencephalogram* (unpublished).

JIRSA, V.K., HAKEN, H.: *A derivation of a macroscopic field theory of the brain from the quasi-microscopic neural dynamics* (submitted to Physica D).

KELSO, J.A.S., BRESSLER, S.L., BUCHANAN, S., DEGUZMAN, G.C., DING, M., FUCHS, A., HOLROYD, T.: "Cooperative and critical phenomena in the human brain revealed by multiple SQUIDs", in: D. DUKE, W. PRITCHARD (Eds.): *Measuring chaos in the human brain*, Singapore (World Scientific) 1991.

PORTUGALI, J.: *Implicate Relations: Society and Space in the Israeli-Palestinian Conflict*, Dordrecht (Kluwer Academic Publishers) 1993.

PORTUGALI, J.: "Inter-representation and cognitive maps", in: J. PORTUGALI (Ed.): *The Construction of Cognitive Maps*, Dordrecht (Kluwer Academic Publishers) in press.

THOMPSON, W.: *On Growth and Form*, London (Cambridge University Press) 1961.

WEIDLICH, W., HAAG, G.: *Concepts and Models of a Quantitative Sociology*, Berlin (Springer) 1983 (= Springer Series in Synergetics, Vol. 14).

Chapter 10

Symbiosis, Mutualism and Cooperation in Biological Evolution*

HANS-DIETER GÖRTZ

I. Introduction

Living organisms interact with their environment, and both positive and negative interactions between organisms are driving forces in the organisation and development of ecosystems. Negative interactions like competition, which is often regarded as the most important of these forces, directly take part in selection. On the other hand, positive interactions in symbiosis, especially mutualism, and cooporation may considerably contribute to the generation of variability and complexity and to an enormous increase in fitnes. This will be elucidated discussing examples of the various types of associations. The evolution of the eucyte (eukaryotic cell) as a basic unit of life and sexual reproduction being unique for eukaryotes has been possible on the basis of symbiosis (interspecific associations) and cooperation (intraspecific associations), respectively.

---

* The author is grateful for critical comments by Dr. M. T. Ghiselin.

HANS-DIETER GÖRTZ

II. Symbiosis

Living together of individuals of different species has been termed symbiosis by De Bary (1879). This definition includes parasitism, commensalism and mutualism - the latter being of mutual advantage for both partners, and it is far from easy to decide, whether an instance of symbiosis is parasitic, commensalistic or mutualistic. Mutualistic as well as parasitic symbioses have influenced biological evolution, and the means of communication between the partners are similar in the different types of symbiosis.

The value of a symbiont must always be regarded in relation to a specific host. Some parasites may harm an intermediate host, but are beneficial for their final host. The tapeworm *Spirometra mansonoides* dwells in the gut of dogs and cats. Larvae hatch from eggs getting into water with the host feces and infect copepods. The larvae of the parasite are further transmitted to mice ingesting infected copepods. Parasitized mice show abnormal growth and change their behavior. Mice infected by the larvae with this or other tapeworms are slower and may be caught more easily by cats or other predators (Mueller 1963). In this and other cases, where the final host does not suffer from the adult tapeworms being small and not seriously harmful, the parasite may rather be beneficial for the final host by increasing the host's chance to catch prey.

The challenge by parasitic and pathogenic infections has led to the development of the various types of immune systems in animals. Competition of the immune systems of the hosts and the defense mechanisms of parasites is driving further evolution of both. The most important parasite of man, *Plasmodium*, causative agent of malaria, lives in the red blood cells, where it finds shelter and is hardly accessable to the immune system. Manifold mechanisms of *Plasmodium* impede the immune system so that the parasites may not be killed when leaving a blood cell to invade another one (Enders 1994).

The causative agent of the sleeping sickness, the protozoon *Trypanosoma*, does not even hide in the blood cells. It swims freely and multiplies in the blood, where it finds plenty of nutrients. However, in the blood it is easily accessible to the immune system of the host, and a dense coat of surface proteins causes the production of great amounts of antibodies that function in the destruction of the parasites. While most of the parasites are killed, few have meanwhile changed their coat expressing a surface protein that is not recognized by the first type of antibody. A second type of antibodies has to be produced, and meanwhile the parasites with the new coat have time to multiply again. When the second antibody has accumulated to attack the protozoa, some of them have again changed their coats (Risse 1994). *Trypanosoma* has more than a hundred genes in its genome for surface proteins. The repetitive cycles of surface protein expression and further means of the parasites compete with the immune system and may finally exhaust it.

Inhabitants of endemic areas of malaria and sleeping sickness suffer from these infections but often are less dramatically harmed than individuals from other parts of the world when infected by *Plasmodium* or *Trypanosoma*. It has been found that certain blood groups favor a partial resistence to malaria or sleeping sickness, but the relatively greater resistence of inhabitants of endemic areas of the parasites is not fully understood. It indicates, however, that a further evolution of our immune system as a reaction to certain pathogens may be possible. Hosts that are well adapted to certain parasites are superior to related species that are poorly adapted and harmed. This advantage explains the selective pressure towards a tolerance to parasites and the development of immune systems.

The ciliate *Paraholosticha sterkii* is an example that the tolerance to an infection that results in a high prevalence of parasites (the percentage of infection) in the populations may be of selective advantage (Görtz/Dieckmann 1980). Local populations of *P. sterkii* are frequently found to be infected by flagellates. The ciliate can multiply, encyst under unfavorable conditions and again excyst with the parasites, while competitive ciliates are killed when infected by the flagellates.

The infection by a parasite or pathogen may sometimes provide the host with an immediate selective advantage within a species. A considerable number of different endosymbiotic bacteria confer killer traits upon their host ciliates (for a review see Heckmann/Görtz 1991). Cells bearing these symbionts may kill cells that do not bear symbionts. These cells are called killers, those who may be killed are called sensitives. The symbiotic bacteria may only be maintained in a host cell with certain allelic conditions for some genes. The ecological significance of symbionts which confer killer traits upon their hosts appears obvious and was proven by Landis (1988), who observed a strong selective disadvantage for sensitives in competition studies at conditions found in nature. The toxin responsable for killing has not been identified, and also the mechanism by which host cells infected with killer symbionts resist being killed is not understood. There are evidences suggesting that the toxin may be a protein, and in some cases it was found that only symbionts infected by bacteriophages contribute to the killer trait. The loss of these bacteriophages resulted in the loss of the killing capacity, but the symbiotic bacteria still provided their host cells with a resistence against the toxic principle.

Symbioses where the symbionts do not dwell within body cavities or even within host cells may nevertheless be intimate and of ecological and evolutionary significance. A number of these are mutualistic such as the cleaning symbioses, the so called trophobiosis of ants and aphids and such widespread symbioses as mycorrhiza and the rhizobium - legume symbioses. Also the pollination symbioses of various flowering plants and their pollinating insects have evolved into intimate associations. In animals various types of behavioral symbioses are found, e.g. the living of mostly insect myrmecophiles in the colonies of ants. Several beetles, larvae of butterflies and other insects are living with ant societies as commensals (sharing the food but not harming their hosts) or parasites. Myrmecophiles have developed a great variety of traits to overcome the defense mechanisms of the ant colonies and they are masked by certain behavioral traits or by their smell or taste. The latter is reminiscent to the masking mechanism of *Schistosoma*, a worm living in the veins of humans. This trematode verm, the causative agent of bilharziosis, adsorbes human blood group antigens and thereby "looks" to the immune system like a huge red blood cell. Like this some myrmicophiles, e.g. the beetle *Atemeles pubicollis*, may deceive their hosts by the release of pheromones that appear identical to those of the ants (Hölldobler/Wilson 1990).

Another trait has been developed in the larvae of lycaenid butterflies that produce sweet secretions being attractive to their host ants. The butterfly larvae are carried into the ant nest, where they participate in the food and even eat larvae of the ants. Most of these myrmecophiles depend upon their hosts at least for certain stages of their life cycle, and their adaptations are often higly evolved and sophisticated (Wilson 1971; Hölldobler/Wilson 1990).

Most fascinating behavioral symbioses are the various types of social parasitism in social hymenoptera: temporary social parasitism, where a young queen may invade the nest of a colony, kills the resident queen to take over the colony; dulosis, that is facultative or obligate slave making; inquilinism finally is a permanent and obligate parasitism, where workerless ants completely depend upon their host colonies. In the latter case the parasite's offspring reared by the host workers are fertile males and females only. The reason of the great number and variability of social parasites (ants, wasps and bees) certainly is that social insects are extremely successful. On the other hand it seems that the highly evolved communication systems (see paragraph *Cooperation* in this article) of social insects offer to a parasite numerous chances for cheating. The principles of communication of host and parasite, including the mechanisms of resistance against invasion and defense mechanisms of the parasite though on a different physiological level are reminiscent to those found in individual parasitism and symbiosis.

III. Mutualism

In this special type of symbiosis both partners benefit from the association, although the degree of advantage may not be equal. Even, mutualism may be labile because host and symbiont are under selection for increased exploitation of the other. An example is the symbiosis of the yucca plant with its pollinator, the yucca moth *Tegeticula yuccasella.* The moth progeny feed on yucca seeds, but consume only small proportions of the developing seeds. Specializations of the moth to collect and transport pollen among yucca flowers and reciprocal specialization of the plant has led to mutual dependence between yuccas and this moth. Neither species can successfully reproduce without the other. Pellmyr et al. (1996) recently presented evidence for reversal of this obligate mutualism. They reported the existence of cheater species derived from obligate pollinators. Their progeny still feed on yucca seeds, but the moths are no longer pollinating the yucca. As long as the cheaters coexist with the pollinators and do not harm the plant too much, they will be tolerated. Otherwise, the evolution of cheating by one partner may lead to reciprocal extinction.

On the other hand, it is not surprising that parasitism has frequently developed into mutualism. Mites associated with insects are mostly parasitic, others are commensals. Springett (1968) studied the nature of the association of the mite *Poecilochirus necrophori* with the burying beetles *Necrophorus investigator* and *N. humator* and found that the beetles in nature commonly carry from 10 to 30 mites, while in laboratory cultures the load of mites was up to 800 / beetle. Nymphs of the mites attack the eggs and small larvae of the fly *Calliphora* spp. *Necrophorus* are unable to utilize corpses of mice and fail to breed in the absence of mites presumalby due to the action of these carnivorous mites. The nymphs of the mites are attracted by the smell of

Necrophorus and even feed on droplets regurgitated by the beetles. While the beetles are unable to breed successfully in the absence of mites, the mites vice versa appear to be unable to breed in the absence of *Necrophorus*. Thus, the relationship between *Necrophorus* beetles and *Poecilochirus* mites has become obligatory.

Another example of a symbiosis having developed from parasitism into mutualism has been described by Jeon (1983) in the protozoon *Amoeba proteus*. He observed bacterial infections in his laboratory cultures. The bacteria were harmful to the host cells for a period of more than a hundred fissions of the amoebae. Thereafter, the amoebae became dependent of the bacteria. Thus, within a relatively short period a pathogenic microorganism became an essential symbiont. This phenomenon turned out to be reproducible when other *A. proteus* cells were infected by the bacteria. While it is known that the bacteria that are infected by a plasmid provide the host cell with a 29 kDa protein, the mechanisms by which the host became dependent has not been discovered.

The bacteria that infected the amoebae probably have been free living contaminants in the culture medium. They have been taken up by the amoebae by phagocytosis being the normal mechanism of feeding in these organisms. There is no way for such ingested microorganisms to escape from the cell again, and most bacteria that are ingested are rapidly killed and digested. Only in rare cases an ingested bacterium resists the killing mechanisms and digestive attacks and then may persist in the host cell.

Infections of animals or protozoa by algae of the genus *Chlorella* are favorable to the host and various species are never found free of these so-called zoochlorellae in nature. The hosts are supplied with carbohydrates -especially maltose - and other metabolites by their symbionts and thus gain considerable amounts of energy from these symbioses (Muscatine et al. 1967; Reisser 1976). The hosts may even show altered behavior. Green paramecia accumulate in the light but cannot do so when free of symbionts. While the hosts certainly take advantage of their algal symbionts, the maintenance of the symbionts may also be seen as due to an inability of the paramecia to digest the algae - probably because of the molecular surface structure of the algae. Even in starvation a host cell cannot digest its symbionts. Aposymbiotic (symbiont-free) *P. bursaria* cannot resist an infection by various bacteria, yeasts or algae, but in infection experiments *Chlorella* was found dominant over other microorganisms (Görtz 1982). The reason why *P. bursaria* is always found infected with *Chlorella* in nature may therefore be, that cells that have become aposymbiotic are readily infected by these algae again.

Whereas the hosts are favored by a symbiosis, the individual symbionts usually are completely dependent once they have become a symbiont. Intracellular algae in paramecium or hydra do not have the means to leave their host cells, even though few of them may be released accidentally. Symbiotic algae depend upon the substrates provided by the host cell and rely on the inability or inactivity of the host cell to digest their symbionts. Although the algae are literally kept as prisoners and their propagation is limited by the host cells, the symbiosis may nevertheless be regarded as mutualistic, because the algae are nurished by their host cell, gain motility (they are not motile on their on) and find shelter.

Like *Paramecium bursaria* and *Hydra viridissima* in fresh water various symbiotic algal symbionts are found in marine organisms as well. Algal symbionts in marine organisms are called zooxanthellae. The symbionts mostly are diatomes or dinoflagellates. Some primitive animals like the cnidarian *Anemonia sulcata* are facultative hosts of zooxanthellae, whereas the turbellarian *Convoluta convoluta* completely depends upon its algae. The same is true for some protozoa. The giant foraminifera living in oligotrophic warm parts of the oceans like in the Caribbean sea take up diatomes or dinoflagellates and are then nurished by their symbionts. The algal symbionts are dwelling in vesicles (perialgal vacuoles) themselves and thus depend upon nurishment by the host cell. These symbioses may therefore be regarded mutualistic.

Organisms bearing zooxanthellae certainly have occupied new niches. Because of these symbioses they are abundant in environments where related organsism free of symbionts can hardly survive. Green foraminifera and green corrals that are nurished by their symbionts (Kremer et al. 1980; Taylor 1983) have considerably contributed to reef formation.

In other associations the interaction may develop into a complete exploitation of the symbiont. This is most obvious when the algae are digested and only the chloroplast kept. One of the best known example of such plastid symbioses is the ciliate *Mesodinium rubrum* that keeps the chloroplast of a chrysophycean alga (Lindholm 1985). Also, the sea slug *Elysia viridis* and some other molluscs (Hinde 1983) only partly digest their algal food and keep the chloroplasts of the alga that remain functionally active.

Dinoflagellates are regarded as originally heterotrophic and colourless. There are numerous indications that chloroplasts of green dinoflagellates derived from secondary symbioses with green eukaryotic protists. In the oceans green dinoflagellates now appear to be the most important photosynthetic organisms. The symbiotic origin of their chloroplast therefore has great ecological impact and has largely influenced biological evolution.

Several intracellular symbioses or endocytobioses (for a discussion of terms see Sitte 1993) of protozoan hosts and bacterial symbionts may be regarded as mutualistic, too. The bacterium *Polynucleobacter necessarius* is an essential symbiont of the ciliate *Euplotes aediculatus*. Other related species of fresh water *Euplotes* also depend upon these symbionts (Heckmann 1983). The endocytobiotic bacteria in *Amoeba proteus* investigated by Jeon (1983) have become essential symbionts, too (see above). Other bacterial symbiont of almost organelle-like nature that clearly contribute to the metabolism of the host cell are found in *Crithidia*, a parasitic flagellate (Chang/Trager 1974). All these cases of essential and obligate bacterial symbionts show that at least in protozoa new cell organelles may evolve by endosymbiosis even nowadays.

Many evidences point to a symbiotic origin of such important cell organelles as mitochondria and chloroplast (Margulis 1981). Still, the discussion is not settled, whether all mitochondria result from one or more symbionts and also whether all chloroplasts are derived from one or more symbionts. An early objection to the hypothesis of an endosymbiotic origin of cell organells was that no intimate endocytobiosis with essential symbionts was known. Now, as reported above, especially in protozoa numerous of such cases have been described.

Even in intimate endocytobioses a gene transfer between the partner may not necessarily take place, but we have good evidence in a number of cases (e.g. Martin et al. 1992; Sitte 1993; Palmer 1995). Such a gene transfer from a symbiont to its host may be regarded as interspecific recombination. As a result of gene transfer and loss of other genes the remaining genome of organelles as well as that of essential symbiotic bacteria is always smaller than that of free living bacteria (Soldo et al. 1983; Schmidt 1982; Sitte 1993), and it is always the symbiont that finally looses its individuality.

IV. Cooperation

The term cooperation in this article will be used for intraspecific positive interactions. Cooperation may already be found among bacteria, e.g. *Lactobacillus acidophilus* (Doederleins bacillus) that acidify their environment colonizing the vaginal mucosa and by acidification prevent competitors to settle there. Certain protozoa release species specific growth factors into the water stimulating growth of each other. Plant exsudates in the soil may impede the growth of competing species and thus have a cooperative effect. For higher plants it was found that neighbors may buffer physical stress of each other, which is most obvious in deserts. Especially for recruitment such a nurse-plant effect appears to be common not only in plants but also in mussels, barnacles and other animals (Bertness/Callaway 1995).

Cooperation often directly aims to increase reproductive success of the cooperating individuals. The implication of cooperation on reproductive success is most obvious in insect societies. Up to several thousand individuals (workers) act in favor of the reproductive success of their nestmates - sometimes a single queen. Workers of the various castes of social insects do no have any own offspring but completely act in favor of the reproduction of near relatives. Although there are only one or few fertile females in an insect colony, the abundance of these insects by far exceeds that of solitary insects.

A morphological, physiological and behavioral diversification of castes that is more pronounced in termites than in hymenoptera is the basis of a sophisticated division of labor in insect societies. But even in the colonies of those hymenopteran species where the caste heteromorphism is small the workers perform a great variety of tasks, such as foraging, brood care, defense and nest construction. The numbers of workers being engaged in specific tasks may change as the situations change according to climate, availability of resources etc. or internal factors such as the physiology of the queen or the age of the individual workers. None of those is capable of understanding the colony organization or present necessities. Rather, the workers may respond to local and transient cues (Hölldobler/Wilson 1990).

The decision of an individual which task to perform is certainly influenced by the rate of interaction with nestmates engaged in like or unlike tasks. Information may not only come from direct communication. E.g. in a wasp colony a forager's decision whether to collect more building material or stop this task depends on how long she had to wait before a builder nestmate accepted her last load (Gordon 1996). Social insects communicate mostly through chemical and tactile cues, and the informations given are manifold and specific. However, whether at all and how an individual will react on a given information depends on its own phyiological condition. E.g. in case of alarm a soldier may proceed towards a disturbance to attack an enemy, while a small worker may take eggs or young larvae, carry them away and try to hide. External stimuli are interpreted (differently) according to the physiological condition of the individual. The ecological and reproductive success of a social insect species certainly depends on the efficacy of task allocation in its colonies, but little is known about the evolution of task allocation (for a review see Gordon 1996).

The social way of life has evolved at least twice favored by different phenomena, i.e. in the hymenoptera and termites. In termites the evolution of eusociality apparently has been favored by the fact that they have to pass on their gut symbionts to newly molted colony members. In the family of hymenoptera - including the ants, wasps and bees - the evolution of sociality has been favored by the unique mechanisms of sex-determination. Individuals that are heterozygous for a certain gene (sex-locus) are female, individuals that are hemizygous or homozygous for this gene are male. Fertilized eggs that are diploid normally develop into females, unfertilized eggs that are haploid normally develop into males (drones). Females may either be workers or queens: The development to either workers or queens is determined trophically. In honey bees the young larvae are determined to develop into queens, when they get predominantly royal jelly (a nutricious secretion of the pharyngeal glands of young worker bees) in the first three days of their life.

The genetic consequences of haplo-diploidy in hymenoptera are that females get half their genes from their mothers and half their genes from their fathers. Drones get all their genes from their mother (they do not have fathers). Sisters share 3 / 4 of genes by descent (provided they have the same father). Thus, sisters are genetically closer related to each other than are mothers to daughters. To explain the origin and evolution of eusociality Hamilton (1964) developed the kin selection theory that may be defined as the alteration of the frequencies of genes shared by relatives through actions that favor or disfavor their relatives' survival and reproduction. Workers will increase their inclusive fitness more by caring for their younger sisters (larvae of workers or new queens) than by investing to an equal degree into own offspring (Hamilton 1964).

Cooperation is also found in protozoa, e.g. the slime mold *Dictyostelium*. Individual amoebae of the slime mold aggregate triggered by cAMP (cyclic adenosine monophosphate) to form a fruiting body. Pairing and conjugation in certain ciliates is initatied by gamones (pheromones), small polypeptides, glycoproteins or in one case a serotonin-like compound that may be recognized by signal-specific receptors (Miyake 1981; Brünen-Nieweler et al.

1994). Intercellular communication in protozoa is reminiscent to that of embryonic cells in ontogenesis of animals, where cells that have basically identical genoms communicate, differentiate and form an organism. As in an insect society cooperating units that originate from unspecialized cells specialize to be capable of performing definite tasks.

The most significant cooperation between individuals certainly is that of males and females in sexual reproduction. Sexual reproduction that includes meiosis, formation of gametes, and fertilization has been invented in early evolution of eukaryotes soon after the evolution of the eukaryon (eukaryotic nucleus) and it must be regarded as one of the main features of eukaryotic organisms. Already most primitive protists show sexual reproduction although they are still able to reproduce vegetatively simply by mitotic cell division. In haploid protists the cells themselves may function as gametes and fuse, thereby fertilizing each other. In some cases both gametes appear identical (at least morphologically); they are then called isogametes. A tendency to generate morphologically different and differently behaving gametes leads to anisogamety, where one of the gametes is smaller and more active than the other. The final type is oogamy, where one gamete may be called oocyte (egg cell) and the other sperm. This is the type of gametes found in higher organisms. Even animals as e.g. tapeworms and many snails that are hermaphrodites show sexual reproduction with both types of gametes. Fertilization generates a diploid nucleus in the egg, whereas meiosis as a special type of nuclear division generates haploid nuclei. Haploid organisms divide by meiosis immediately after fertilization, whereas in diploid organisms (most animals and plants) fertilization is found immediately after meisosis.

A result of sexual reproduction is the evolution of eukaryotic genomes and the generation of a variability of organisms that could never have been obtained in prokaryotes with their mechanisms of recombination being comparably much less effective. The rank of sexual reproduction in the evolution of eukaryotes is also emphasized by the mechanisms of its inheritance. Sex (male or female) is the only phenotype that may be inherited by special chromosomes (notwithstanding that there are genes on the chromosomes respronsible for sex determination); many higher eukaryotes have such sex chromosomes (heterosomes). The mode of chromosome segregation in meiosis ensures that a sex ratio of 1 : 1 is maintained in many species.

V. Conclusion

Living organisms are not designed *a priori* with regard to a certain physiological or ecological function. Rather, variability is brought about by change, recombination and extension of the genoms of existing organisms. A further mechanism contributing to increase variability is the association or fusion of organisms. Symbiosis including mutualism accidentally brings together two complete genomes of different species that have evolved independently. Namely in case of mutualism the genoms of the partners may partly merge and in the process of further evolution may be reduced by mechanisms of gene-transfer and economizing the genome. The processes connected with the coordinated function or merging of genomes of host and symbiont must be regarded as a special type of interspecific recombination. In comparison to mutations and to recombination in meiosis (in diploid organisms) the effects of symbiosis on the phenotypes of symbiont and host may be immediate and drastic. Not only may symbiosis and cooperation help to overcome the lack of certain abilities of the individal partners and thus enable them to colonize new ecological niches, more important is that symbiosis and cooperation obviously have a synergetic effect having repeatedly led to a quantum jump in evolution, e.g. in the evolution of the eukaryotic cell, in the evolution of sexual reproduction and of multicellular organisms.

The positive effect of cooperation of genetically and phenotypically identical or similar individuals may primarily depend upon the number of partners like in the nurse-plant example. However, the generation of heterogeneous phenotypes (in an insect society; male and female gametes; differentiated cells in a metazoon) in a second step offeres new possibilities. The cooperation of phenotypically different cells / individuals that may individually be allocated specific tasks is of great advantage.

Sexual reproduction depending on the cooperation of male and female gametes resulted in a higher speed of evolution and an immense variety among eukaryotes. This success of sexual reproduction is evidently shown by the much greater variability of eukaryotes compared with prokaryotes.

Whereas male and female are genetically different at least in most higher animals, a genetical difference is not absolutely necessary to generate different phenotypes for cooperation. The castes of social insects are phenotypically different but genetically identical. Cooperation of individuals with different abilities have made social insects the predominant group of animals next to man. In contrast to symbiosis, where the genomes of partners may tend to merge, in cooperation the phenotypic difference and individuality of partners in cooperation is of advantage and was maintained or became even more pronounced during evolution. But like in symbiosis the success of cooperation results from the synergetic behaviour of individuals that in this case have identical genotypes but different abilities. The same phenomenon is found in metazoa where hundreds to millions of cells cooperate in an individual. In ontogenesis cells derived from the egg differentiate and develop to perform specialized metabolic tasks in a well organized physiology. The final result again is due to the synergetic effect of innumerable specialized cells communicating and cooperating by manifold systems.

References

BERTNESS, M.D., CALLAWAY, R.: "Positive interactions in communities", *Trends in Ecology and Evolution*, 9 (1994), pp. 191-193.

BRÜNEN-NIEWELER, C., MEYER, F., HECKMANN, K.: "Expression of the pheromone 3-encoding gene of *Euplotes octocarinatus* using a novel bacterial secretion vector", *Gene*, 150 (1994), pp. 187-194.

CHANG, K.-T., TRAGER, W.: "Nutritional significance of symbiotic bacteria in two species of hemoflagellates", *Science*, 183 (1974), 531-532.

DE BARY, A.: *Die Erscheinungen der Symbiose* (The Manifestations of Symbiosis), Strassburg (Trübner) 1879.

ENDERS, B.: "Malaria - eine globale Herausforderung", in: K. HAUSMANN, B. P. KREMER (Eds.): *Extremophile (Extremophils)*, Weinheim (VCH) 1994, pp. 381-408.

GÖRTZ, H.-D.: "Infection of *Paramecium bursaria* with bacteria and yeasts", *Journal of Cell Science*, 58 (1982), pp. 445-453.

GÖRTZ, H.-D.: "Endonucleobiosis in ciliates", *International Review of Cytolology*, 102 (1986), pp. 169-213.

GÖRTZ, H.-D., DIECKMANN, J.: "*Leptomonas ciliatorum* n.sp. in the macronucleus of a hypotrich ciliate", *Journal of Protozoology*, 34 (1987), pp. 259-265.

GORDON, D. M.: "The organization of work in social insect colonies", *Nature*, 380 (1996), pp. 121-124.

HAMILTON, W. D.: "The genetic evolution of social behaviour, I, II.", *Journal of Theoretical Biology*, 7 (1964), pp. 1-52.

HECKMANN, K.: "Endosymbionts of Euplotes", in: K.W. JEON (Ed.): *Intracellular symbiosis.*, New York (1983) pp. 111-144.

HECKMANN, K., GÖRTZ, H.-D.: "Prokaryotic symbionts of ciliates", in: A. BALOWS, H.G. TRÜPER, M. DWORKIN, W. HARDER, K. H. SCHLEIFER (Eds.): *The Prokaryotes,* 2nd ed., Berlin, Heidelberg, New York (Springer) 1991, pp. 3865-3890.

HINDE, R.: "Retention of algal chloroplasts by molluscs", in: L. J. GOFF (Ed.): *Algal Symbiosis. A continuum of interaction strategies*, Cambridge (Cambridge University Press) 1983, pp. 97-198.

HÖLLDOBLER B., WILSON, E. O.: *The Ants,* Berlin (Springer) 1990.

JEON, K. W.: "The large, free-living amoebae: wonderful cells for biological studies", *Journal of Eukaryotic Microbiology,* 42 (1995) pp. 1-7.

KREMER, B. P., SCHMALJOHANN, R., RÖTTGER, R.: "Features and nutritional significance of photosynthates produced by unicellular algae symbiotic with larger foraminifera", *Marine Ecology Progress Series*, 2 (1980), pp. 225-228.

LANDIS, W.G.: "Ecology" in: H.-D. GÖRTZ (Ed): *Paramecium*, Berlin (Springer) 1988, pp. 419-436.

LINDHOLM, T.: "*Mesodinium rubrum* - a unique photosynthetic ciliate", in: H.W. JANHASCH, P. J. WILLIAMS (Eds.): *Advances in Aquatic Microbiology,* Vol. 3 (1985), pp. 1-48.

MARGULIS, L.: *Symbiosis in Cell Evolution*, San Francisco (Freemann) 1981.

MARTIN, W., SOMERVILLE, C. C., LOISEAU-DE GOER, S.: "Molecular phylogenies of plastid origins and algal evolution", *Journal of Molecular Ecology*, 35 (1992), pp. 385-404.

MIYAKE, A.: "Physiology and biochemistry of conjugation in ciliates", in: M. LEVANDOWSKY, S.H. HUTNER (Eds.): *Biochemistry and physiology of protozoa*, Vol. 4, New York (Acad. Press), 1981, pp. 125-198.

MUELLER, J.F.: "Parasite-induced weight gain in mice.", *Annals of the New York Academy of Sciences*, 113 (1963), 217-233.

MUSCATINE, L., KARAKASHIAN, S., KARAKASHIAN, M.: "Soluble extracellular products of algae symbiotic with a ciliate, a sponge and a mutant hydra", *Comparative Biochemistry and Physiology*, 20 (1967), pp. 1-12.

PALMER, J. D.: "Rubisco rules fall; gene transfer triumphs", *BioEssays*, 17 (1995, pp. 1005-1008.

PELLMYR, O., LEEBENS-MACK, J., HUTH, C. J.: "Non-mutualistic yucca moths and their evolutionary consequences", *Nature*, 380 (1996), pp. 155-156.

REISSER, W.: "Die stoffwechselphysiologischen Beziehungen zwischen Paramecium bursaria Ehrbg. und Chlorella spec., in der Paramecium bursaria-Symbiose. II. Symbiose-spezifische Merkmale der Stoffwechselphysiologie und der Cytologie des Symbioseverbandes und ihre Regulation", *Archiv für Microbiologie*, 111 (1976), pp. 161-179.

RISSE, H.-J.: "Afrikanische Trypanosomen - Meister im Austricksen des Immunsystems", in: K. HAUSMANN, B. P. KREMER (Eds.): *Extremophile (Extremophils)*, Weinheim (VCH) 1994, pp. 361-380.

SCHMIDT, H. J.: "Isolation of omikron-endosymbionts from mass cultures of *Euplotes aediculatus* and characterization of their DNA", *Experimental Cell Research*, 140 (1982), pp. 417-425.

SITTE, P.: "Symbiogenetic evolution of complex cells and complex plastids", *European Journal of Protistology*, 29 (1993), pp. 131-143.

SOLDO, A. T., BRICKSON, S.A., LARIN, F.: "The size and structure of the DNA genome of symbiont particles in the ciliate *Parauronema acutum. Journal of General Microbiology*, 129 (1983), pp. 1317-1325.

SPRINGETT, B. P.: "Aspects of the relationship between burying beetles, *Necrophorus* spp. and the mite Poecilochirus necrophori Vitz", *Journal of Animal Ecology*, 37 (1968), pp. 417-424.

TAYLOR, K.L.: "The coral-algal symbiosis", in: L.J. GOFF (Ed.): *Algal Symbiosis. A continuum of interaction strategies*, Cambridge (Cambridge University Press) 1983, pp. 19-36.

WILSON, E.O.: *The insect societies*, Cambridge, Mass. (Belknap Press of Harvard University Press) 1971.

WILSON, E.O.: *Sociobiology, the new synthesis*, Cambridge, Mass. (Belknap Press of Harvard University Press) 1975.

Part V

Evolutionary Biology and Economic Theory

Chapter 11

Are Economic Systems Like Organisms?*

MAE-WAN HO

I. The Need for an *Organic* Theory of the Organism and of Economic Systems

Hodgson (1993) and Ormerod (1994) are the latest among a string of economists to declare their own discipline "in crisis" within the past 15

* I am grateful to Ulrich Witt, Bruce Weber, Susan Oyama, Peter Saunders and Peter Koslowski for comments on earlier drafts of this paper. In addition, I would like to thank Teddy Goldsmith, Vandana Shiva, Martin Khor and many other colleagues of the Third World Network for raising my awareness on world economic issues. None of those mentioned should be held responsible for the shortcomings.

years. And like the others before them, they trace the crisis to the mechanistic foundations of modern western science itself. They call for an alternative, organicist approach to economics. Precisely the same critique of neo(classical) Darwinian theory of evolution has been taking place since the 1970s (see Saunders, and Oyama, this volume; Ho/Saunders 1984; Ho/Fox, 1988; Ho 1996a), with a "new organicism" emerging (see Ho 1996b and references therein) which explicitly affirms Whitehead's (1925) view that nature cannot be understood except in terms of a theory of the organism that *participates* in knowing and in constructing reality. This happy coincidence in the evolution of ideas entices me to explore more tangible links between a tentative theory of the organism and sustainable economic systems.

As Hodgson and Ormerod and many others have observed, Darwinian and free-market theories are of the same cloth. The ideology of unbridled competition and self-interest, supposed to govern supply and demand in the free market, to a large extent, inspired Darwin's theory of evolution by natural selection - that unbridled competition and self-interest lead inevitably to the survival of the fittest. Free-market theory developed in the wake of industrialization and the rise of capitalism, and became translated into mathematics in analogy with the laws of equilibrium thermodynamics and statistical mechanics. Neo-Darwinian theory, expressed mathematically in population and biometrical genetics, is based on the same equilibrium, mechanistic assumptions, and has even closer links with statistics (see Weber/Depew, this volume). The common roots of human economy and the "economy of nature" has been made explicit by Ghiselin (1974).

Koslowski (1996) has recently produced a comprehensive critique of both sociobiology and neo-Darwinian bioeconomics, pointing out that,

"The adoption of sociobiological, evolutionary theories in our culture is not merely a scientific question as to the correctness of hypothesis, but also a normative question of social philosophy, whether we want to and ought to understand ourselves by the sociobiological model ... The way a society defines itself simultaneously constitutes an aspect of this society's reality ..." (pp. 83-84).

The present paper offers a theory of the organism and of a sustainable economic system as an organism. I hope thereby to define an alternative social reality to the neo-Darwinian, which many of us would find more consonant with our deepest experience of nature's unity.

A sustainable system has all the essential characteristics of an organism - an irreducible whole that develops, maintains and reproduces, or renews, itself by mobilizing material and energy captured from the environment. What is the nature of the material and energy mobilization that makes an organism? I begin with a brief description of a tentative theory of the organism - developed in detail elsewhere (Ho 1993; 1994a; 1995a,b; 1996b,c) - as a dynamically and energetically *closed* domain of cyclic non-dissipative processes coupled to irreversible dissipative processes, which effectively frees the organism from thermodynamic constraints so that it is poised for rapid, specific intercommunication, enabling it to function as a coherent whole. I shall then show how this novel theoretical framework may begin to provide *normative* criteria for sustainable economic systems, thereby also exposing some of the inadequacies of current models and assumptions.

II. What Is an Organism?

1. The Organism Transcends Thermodynamic Constraints

Organisms are so enigmatic from the physical, thermodynamic point of view that Lord Kelvin, co-inventor of the second law of thermodynamics, specifically excluded them from its dominion (Ehrenberg 1967). As distinct from heat engines which require constant input of heat to do work, organisms are able to work without a constant energy supply, and moreover, can mobilize energy *at will*, whenever and wherever required, and in a perfectly coordinated way. Similarly, Schrödinger (1944) was impressed with the ability of organisms to develop and evolve as a coherent *whole*, and in the direction of increasing organization, in defiance of the second law. He suggested that they feed upon "negative entropy" to free themselves from all the entropy they cannot help producing. The intuition of both physicists is that energy and living organization are intimately linked.

The idea that open systems can "self-organize" under energy flow became more concrete in the concept of *dissipative structures* developed by Prigogine, (1967) and Haken (1977) that depend on the flow and dissipation of energy, such as the Bénard convection cells and the laser. In both cases, energy input results in a phase transition to global dynamic order in which *all* the molecules or atoms in the system move coherently. From these and other considerations, I have identified Schrödinger's "negative entropy" as "stored mobilizable energy in a space-time structured system" (Ho 1993, 1994a, 1995b), which begins to offer a possible solution to the enigma of living organization.

2. The Organism Stores Mobilizable Energy Over all Space-Times

The key to understanding the thermodynamics of the living system is neither energy flow nor energy dissipation, but energy storage under energy flow (Fig. 1). Energy flow is of no consequence unless the energy is trapped and stored within the system where it circulates before being dissipated. A reproducing life cycle, i.e., an organism, arises when the loop of circulating energy closes. At that point, we have a life cycle within which the stored energy is mobilized, *remaining stored as it is mobilized*, and coupled to the energy flow.

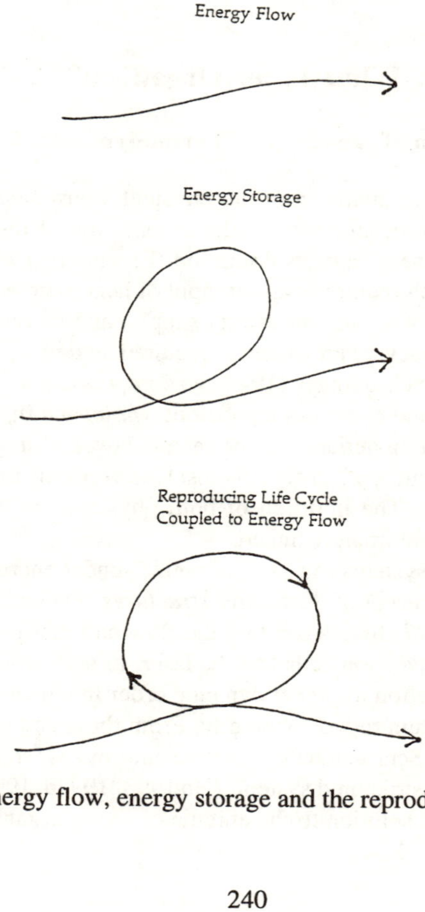

Figure 1. Energy flow, energy storage and the reproducing life-cycle

Energy storage depends on the highly differentiated space-time structure of the life cycle, whose predominant modes of activities are themselves cycles of different sizes, spanning many order of magnitudes of space-times, which are all coupled together, and feeding off the one-way energy flow (Ho 1993; 1995b; 1996c,d). The more coupled cycles there are in the system, the more energy is stored, and the longer it takes for the energy to dissipate. The average *residence time* of energy in the system (Morowitz 1968) is therefore a measure of the *organized complexity* of the system. An intuitive representation is given in Figure 2. I have proposed (Ho 1994a; 1995b) that open systems capable of storing energy tends to evolve towards an extremum, or end-state, in which all space-time modes become equally populated with energy under energy flow. This implies an evolution towards increasing complexity, which we shall come back to later.

Figure 2. The many-fold cycles of life coupled to energy flow

Coupled processes are familiar in metabolism in living systems: practically all thermodynamically uphill reactions - those requiring energy input - are coupled to the thermodynamically downhill ones - those yielding energy - (see Harold 1988; Ho 1995a). That living processes are organized in cycles is also intuitively obvious by a casual examination of the metabolic chart depicting the known biochemical reactions in our body, which shows how the cycles are entangled in a complicated web-like network. Another prominent way in which cycles appear is in the familiar form of the wide spectrum of biological rhythms - with periods ranging from milliseconds for electrical activities of single cells to circadian and circa-annual cycles in whole organisms and populations of organisms (Breithaupt 1989; Ho 1993). These cycles also interlock to give the organism a complex, multidimensional, entangled space-time, very far removed from the simple, linear space and time of Newtonian physics (Ho 1993; 1994b). Thus, integral relationships are maintained between well-known cycles such as the heart beat and the respiratory rate (Breithaupt 1989). Remarkably, mutations in two genes of *Drosophila* which speed up, slow down or abolish circadian rhythm, also cause correlated changes in the millisecond wing beat cycle of the male fly's love song (see Zeng *et al* 1996). This correlation spans seven orders of magnitude of characteristic timescales, reflecting the full extent of *storage and mobilization* of energy in the living system. As all the space-time modes are coupled together, energy input into any mode can be readily delocalized over all modes, and conversely, energy from all modes can become concentrated into any mode. In other words, energy coupling in the living system is *symmetrical,* as argued in detail elsewhere (Ho 1993; 1994a; 1995a,b; 1996c,d).

In analogy with the treatment of the steady-state by Onsager, Helmholtz and others (see Denbigh 1951), I propose (Ho 1966c,d) that *the organism is a superposition of non-dissipative cyclic processes, for which the net entropy production balances out to zero, i.e.,* $\Sigma\Delta S = 0$*, and dissipative, irreversible processes, for which the entropy production is greater than zero, i.e.,* $\Sigma\Delta S > 0$ (Fig. 3).

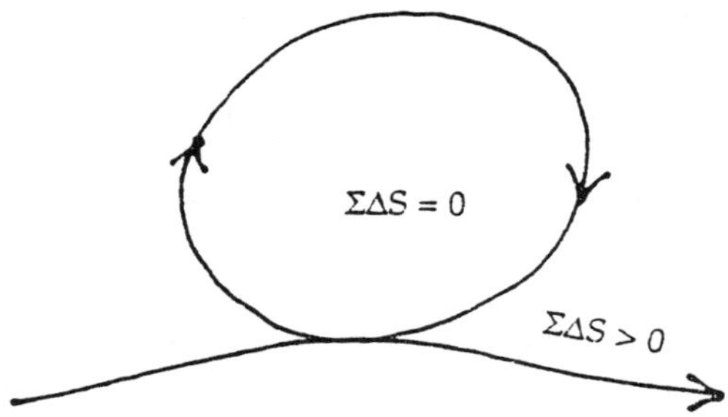

Figure 3. The organism frees itself from the constraints of energy conservation and the second law of thermodynamics.

The cyclic non-dissipative branch will include most living processes because of the ubiquity of coupled cycles, for which the *net* entropy production most probably does balance out to zero, as Schrödinger (1944) had surmised. In this way, the organism achieves *dynamic closure* to become a self-sufficient energetic domain (Ho 1996c,d). The dynamic closure of the living system has a number of important consequences. First and foremost, it frees the organism from the immediate constraints of energy conservation - the first law - as well as the second law of thermodynamics, thus offering a solution to the enigma of the organism posed by Lord Kelvin and Schrödinger. *There is always energy available within the system, for it is stored and mobilized at close to maximum efficiency over all space-time domains.*

Two other consequences of dynamic closure are that, it frees the organism from mechanistic constraints, and creates, at least, some of the basic conditions for *quantum coherence*.

3. The Organism Is Free From Mechanistic Constraints

One of the hallmarks of an organism is its exquisite sensitivity to specific, weak signals. For example, the eye can detect single photons falling on the retina, where the light sensitive cell sends out an action potential that represents a million-fold amplification of the energy in the photon. Similarly, a few molecules of pheromones in the air are sufficient to attract male insects to their mates. No part of the system has to be pushed or pulled into action, nor be subjected to mechanical regulation and control. Instead, coordinated action of all the parts involves rapid *intercommunication* throughout the system. The organism is a system of excitable cells and tissues poised to respond specifically and disproportionately to weak signals, *because the large amount of energy stored can automatically amplify weak signals, often into macroscopic actions* (Ho 1996c,d). That is why organisms cannot be understood in mechanistic terms.

Stored energy is *coherent* energy capable of doing work. That implies the organism is a highly coherent domain, possessing a full range of coherence times and coherence volumes of energy storage. In the ideal, it can be regarded as a quantum superposition of activities - organized according to their characteristic space-times - each itself coherent, so that it can couple coherently, i.e., non-dissipatively, to the rest. The theoretical arguments and empirical evidence for quantum coherence are presented in detail elsewhere (Ho 1993; 1995b; 1996 c,d).

The main implication of quantum coherence for living organization is that it maximizes *both* local freedom and global intercommunication. The organism is in a very real sense completely free (Ho 1996b). Nothing is in control, and yet everything is in control. An organic whole is an *entangled* whole, where part and whole, global and local are so thoroughly implicated as to be indistinguishable, and where each part is as much in control as it is sensitive and responsive. There is no choreographer orchestrating the dance of molecules in the living system. Ultimately, choreographer and dancer are one and the same (Ho 1995a).

III. What Would an Organic Economic System Be Like?

1. The Organic Approach Versus the Neo-classical Mechanistic Approach

Does the theoretical framework of the organism just presented have any relevance for economic systems? While both Hodgson (1993) and Ormerod (1994) stress the need to go beyond the mechanistic, linear approach of the classical theory to a nonlinear, organic approach, there are immediate problems that have to be addressed. First, there is no compelling *a priori* reason to believe that economic systems should be like organisms. Economic systems are constructed by human beings, and experience tells us that as many human institutions are run in hierarchical mechanistic fashion as those that are organic and cooperative.

There is a large element of self-fulfilling prophecy about human action, given that reality *is* shaped by human volition. Ormerod cites some revealing experiments carried out in Cornell University with economics students and students in other subjects. In the first experiment, they were paired up as 'allocator' and 'receiver'. The allocator was given $10 and told to divide the money between the two. The receiver could accept or reject what was offered by the allocator. If the receiver rejected the offer, then neither player received anything. The results showed that economics students, presumably having been taught the importance of self-interest, which, in the west, has a tendency to be confused with *selfishness* (see Ho 1996e for detailed argument that they self-interest and selfishness are distinct), performed significantly more selfishly. In the second experiment, the students were each given some money and asked to divide it into two accounts, one public and the other private. Once all the decisions were taken, the students were told that the money in the public account would be increased by the organizer and would then be equally distributed among the students. For the group of students as a whole, the best solution was to put all their money in the public account, but for the individual, the best strategy was to put all the money in the private account, and still receive a share of the public money. The economics students were found to have contributed on average 20% to the public account, whereas the non-economics students contributed no less than 50%. This is a clear demonstration, not only that there is nothing 'pure' about knowledge in the sense of it being divorced from life (c.f. Hodgson 1993). On the contrary, the wrong kind of knowledge can lead us astray, or in any case, lead us where we do not want to go. Ormerod (1994) and Hutton (1995) both argue that monetarist doctrines have shaped (and ruined) the economies of many nations since the 1970s.

The second problem with the organic theory of economics is that even if the economic system is like an organism, it may be more like a sick organism than a healthy, sustainable one. So, while Ormerod's analysis of unemployment data (in terms of Lotka-Volterra equations of ecology) makes much more sense than the linear classical approach, it leaves us in want of definite criteria whereby one may distinguish a healthy economy from the unhealthy. Here is where I believe a concept of a healthy organism may begin to provide *normative*, diagnostic criteria for a thriving, sustainable economy. In other words, I am proposing that a healthy economic system *should* be like a healthy organism. It is all too easy to forget that an economic system is a society of people bonded by social contract to make their living together, by utilizing and transforming resources, the purpose of which is to achieve a good life for all. It is therefore in everyone's interest to have a healthy economy.

The economy is, to first approximation, an open system through which resources extracted from the 'source' - the ecological environment - flow to a 'sink' - the most immediate mental picture of which is the municipal dump. Models of economic systems as dissipative structures have already been entertained by economists (see for example, Mayntz 1992; Witt 1996). Dissipation or wastage can come in many forms, as we shall see. Various commodities and services are exchanged or traded between 'source' and 'sink', and 'values' are added in processes of manufacture, creative acts of art or artesanship, whose equivalence to energy or otherwise need to be fully justified and explicated, along with such qualities of life as happiness, health, contentment and well-being, not to mention clean air, nutritious food, comfortable shelter from inclement weather and unpolluted environment.

Like an organism, the economic system may be conceptualized in terms of cyclic, nondissipative exchanges or transformations of resources, coupled to the dissipative flows or wastage due to deaths, depreciations and other entropy-generating, irreversible processes. As resources come ultimately from the ecological environment, it makes sense to embed the economic system properly in its ecological setting (Fig. 4).

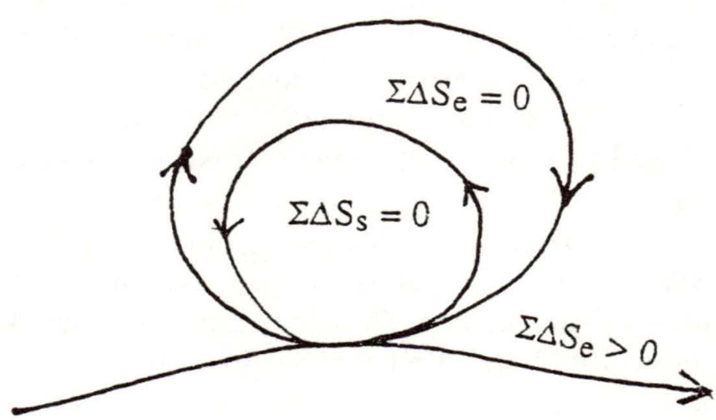

Figure 4. The coupled flows of the economic and ecological cycles in a sustainable economic system.

Figure 4 makes clear that the ecological environment is also conceptualized as a self-sustaining organic system of cyclic non-dissipative processes coupled to the dissipative, one-way energy and material flow. To what extent is that justified? Lovelock's (1979; 1996) Gaia hypothesis proposes that the entire earth is a self-organizing, self-regulating system maintained far from thermodynamic equilibrium under energy flow. In those respects it is indeed like an organism (see also Saunders 1994). (The most conspicuous sign of the earth's self-regulating property is the constancy of its atmosphere, which is a highly non-equilibrium mixture of gases. The atmospheres of Mars and Venus, by contrast, are equilibrium mixtures of spent, or exhaust gases reflecting their lifelessness.) The local environment of an economic system, say, Britain, also has its own local self-organizing, self-regulating properties to some extent, although it is clear that local economies (and environments) are coupled to the global through imports and exports of materials, human beings and capital.

In the context of the planetary ecosystem, it is recognized that human activity has had a far from benign effect. This has prompted the establishment of a "Geophysiological Society" for the study of planetary health (Kump 1996). I suggest that our present model of the organism may begin to offer some diagnostic criteria of health.

2. How to Be a Healthy, Organic Economic System

To be a healthy organic economic system as much as a healthy organism, one should maximize balanced (symmetrical) flows and minimize dissipation for a given rate of inflow of resources. An obvious way to decrease dissipation is to minimize wastage and to recycle resources. Recycling also increases the dynamic closure which is a pre-requisite to a self-sustaining, self-reproducing organism or economy.

Another important factor that decreases dissipation is the degree of space-time differentiation which stores energy and delays the dissipation of the (energetic) value of resources (see Fig. 2). Space-time differentiation, I believe, is the reason complex ecological systems are more stable and viable (see Pimm 1991; DeAngelis 1992). Conversely, it is why intensive farming involving large-scale monocultures have such devastating ecological effects, as they wipe out space-time differentiation (or biodiversity) both directly through clearing vegetation for agriculture and indirectly through harmful effects of herbicides, pesticides and fertilizers on indigenous species. The real importance of biodiversity may be that a diverse ecosystem enables space-time differentiation to be maximized, rather than the number of species *per se*. Of course, the two are expected to be highly correlated, but space-time structure of ecosystems in terms of the life cycle times and spatial distribution of species has more to do with 'niche partitioning' for the most efficient utilization of resources than any conventional explanation in terms of natural selection.

I have proposed that open systems capable of storing energy will tend to evolve towards an increase in space-time differentiation for storing mobilizable energy over all space-times (Ho 1994a), an increase in organized complexity, in other words. This self-organizing principle is manifest in the increasingly complex differentiation of multicellular organisms both in development and in evolution, into organs, tissues and cell types. These form so many nested dynamic compartments and microcompartments down to the interior of cells, facilitating the rapid and efficient mobilization of energy and resources (Ho 1995a).

Some evidence that this principle may apply in ecology has come to my attention recently. There does seem to be an increasing complexity of trophic webs and diversity of niches and microniches in long established ecosystems compared with new ones or ones that have been stressed (Schneider/Kay 1994). The increase in complexity of trophic webs is, moreover, associated with an increase in efficiency of energy and resource utilization. We are reanalyzing the data to see how they bear out the concept of energy storage presented here more precisely (Schneider/Kay/Ho 1996).

In economic systems, the principle of increase in organized complexity can be seen to operate to some extent in the 'division of labour' subsequent to the industrial revolution. Witt (1996) offers a stylized record of the increasing differentiation of labour that accompanies the growth of modern economies. The substitution of human physical work by non-human energy sources and machines gave rise to a whole variety of "lower mental work" and services in controlling, monitoring, tooling and adminstering machinery. The production processes themselves also become more dependent on services as well as explicit technical knowledge and "higher mental work" of all specializations. This in turn, generated the need for training and education. Recent advances in automation and information technology are substituting out some categories and creating others of "lower mental work" such as microchip manufacture and assembly. They also give rise to "higher mental work" in software and hardware design.

There may indeed have been an increase in space-time differentiation of economic systems subsequent to industrialization. Agrarian communities are organized around the village and the fundamental annual crop cycle, with lower and higher harmonics of the yearly cycle of human activities in tune with the rotations and revolutions of planets and stars. By contrast, industrial societies do not have a fundamental timescale. Manufacturing outputs are in hours or days, and may take place simultaneously in spatially distant sites. Fixed capital machinery have typical turnover times of 5 to 10 years whereas human capital (Becker 1966) takes 25 years or more to 'accumulate' and become obsolete in 65 years on average. Firms and small companies have a life time of perhaps 10 to 20 years, while large, multinational corporations may last over several human life-times. Electromagnetic and electronic communications and transactions, on the other hand, can take place across the globe in a matter of split-seconds. To facilitate circulation of resources over the entire gamut of space-times, the financial or banking systems have also evolved to increase space-time differentiation in the availability of capital.

Tapping non-human energy sources and substituting human physical work by machines in industrial societies have no doubt led to an increase in energy storage capacity in proportion to the space-time differentiation of the system. However, the increase in intensity of energy flow and exploitation of environmental resources may be far in excess of any increase in the system's storage capacity. For a given space-time structure, there is a probably an optimum to the rate of in-flux of resources, such that increasing the rate of extraction of resources from the environment beyond that point will no longer make the system more 'energetic', because there is a limit to the amount that can be stored. If resource flow is too fast, it will merely over-heat the system, i.e., increase the rate of dissipation. Whereas natural systems, including indigenous non-intensive agrarian societies have co-evolved with their ecological environment, industrialized systems have generally upset the balance by increasing the rate of exploitation, particularly of non-renewable environmental resources, far beyond the regenerative capacities of the environment, and outstripping the rate at which space-time differentiation can increase to absorb and store the extra inflow of resources.

Thus, while development and industrialization may improve the rate or efficiency of resource extraction, they do not necessarily make the system richer. Increase in resource extraction can indeed impoverish the system as a whole by destroying the environment or even undermine the pre-existing space-time structure of the economic system. This thermodynamic limit to productive resource utilization is the basis of the law of diminishing returns. A case in point is intensive agriculture associated with the Green Revolution in India, which not only depleted ecological biodiversity, but has also thrown small farmers and local distributors out of work, creating conflict that completely undermined the pre-existing socioeconomic structure (see Shiva 1991). Space-time differentiation of both ecosystems and economic systems is an area which will benefit greatly from future investigations.

To be a truly healthy organism, there must be a *balanced* flow of energy and resources. That means maximizing *coupled* flows that are as equal or symmetrical as possible, so that resources can be readily mobilized or distributed throughout the system. In practice, it requires resources from parts of the system in surplus to be promptly diverted to other parts in deficit, as would be achieved with a responsive and responsible banking system. Reciprocity in coupling ensures that the direction of flow may be reversed at other times: debtors and creditors can reverse roles as the need arises. Symmetrical relationship implies the maximization of intercommunication, and *vice versa*.

To illustrate the principles of a healthy economic system, I shall briefly examine how the economy and economic systems may be analyzed with the help of this novel theoretical framework.

3. Money, Energy and Entropy

The first thing that comes to mind wherever the word 'economics' is mentioned is money. It is indeed money that makes the economists' world go round. So, it is all too easy to equate the circulation of money in real world economy with energy in the living system. It has even been fashionable, in biochemistry text-books, to regard the universal energy transduction intermediate, ATP, as 'energy currency'. However, money is by no means equivalent to energy. If one wants a really good analogue to energy in real world economy, it is affection, trust and good will, for which, originally, the gift was a token. Later on, people traded goods or services, value for value, which again depend on trust and goodwill. The problem arose with the introduction of money, which is only arbitrarily related to the real value of things and services. I say this inspite of learned volumes which have been written on the subject (see Crowther 1940).

The result is that all money is not equal. The flow of money can be associated with exchanges of real value or it can be associated with sheer wastage or dissipation. In the former case, it is more like energy flow, in the latter case, it is pure entropy. Classical economists beginning with Adam Smith have devoted much effort towards a theory of "natural value" of things and services (see Barber 1967), on the explicit recognition that there is a natural price at which a 'freely' competitive system may come into equilibrium. However, as the experience of the past centuries has shown, prices are subject to all kinds of manipulations by human beings, which is why the economy cannot be tuned simply by controlling money supply (Hutton 1995). Obviously, the energy/entropy divide is not clear-cut, and there is a continuum from a 'just' and completely equal exchange which is purely energetic, through various degrees of entropic costs when value for money is unbalanced, being either too high or too low, to the purely entropic when the flow is associated with wasteful, unnecessary consumer goods or services, or is essentially decoupled from anything else, for example, in the huge profits reaped in financial and money markets (c.f. Hutton 1995).

Because the economic system depends ultimately on the flow of resources from the natural environment, which has its 'natural ecological economy', entropic costs can either be incurred in the economic system itself, or in the ecosystem. Thus, when the cost of valuable (non-renewable) ecological resources consumed or destroyed are not properly taken into account, the entropic burden falls on the ecological environment rather than on the economic system. But, as the economic system is necessarily coupled to and dependent on input from the ecosystem (see Fig. 4), the entropic burden in the latter will feedback on the economic system as diminished input, so the economic system becomes poorer as a result. In our model, poverty is absolute, as there is a finite optimum rate at which resources can be utilized and transformed. That also means when individuals amass excessive resources, others become poorer in absolute terms. Poverty threatens the survival of the system as a whole. It represents an unbalanced, uncoupled system (see below).

One might think that 'value' and 'worth' are too subjective to say anything rigorous about, but evidence that money is both energetic and entropic is provided by the well-known unreliability of assessing the real "wealth of nations" - equivalent to the mobilizable energy or resources stored in our model - from the Gross National Product (GNP) - the amount of spending, in

US dollars, on goods and services carried out by various sectors of the economy. Several attempts have been made to improve on GNP. One example is the Measure of Economic Welfare (MEW), which takes account of unpaid household work, ascribes value to leisure and costs (negative GNP) to aspects of urbanization such as the necessity to pay for travel to work. Between 1929 and 1965 in the United States, MEW grew, on average, by 1.1% per annum compared to 1.7% per annum in GNP. Another recent suggestion is the Index of Sustainable Economic Welfare (ISEW), which makes deductions for depletion of non-renewable resources and long term environmental damage. For the years between 1950 and 1986 in the United States, ISEW grew by a mere 0.9% per annum compared to 2% in GNP (see Ormerod 1994). These differences are perhaps minimum estimates of the entropic burden borne by the economic system and the ecosystem to which it is coupled. Entropic burdens, if unrelieved, are 'deaths' of the system.

Another suggestive estimate of energetic yield versus entropic costs come from the comparison of 25 rice cultivation systems (see Shiva 1991), of which 8 are *pre-industrial* in terms of low fossil fuel input (2-4%) and high labour input (35-78%); 10 are *semi-industrial* with moderate to high fossil fuel input (23 - 93%) and low to moderate labour input (4 - 46%); and 7 are *full-industrial* with 95% fossil fuel input and extremely low labour input of 0.04 to 0.2%. The total output per hectare, calculated in GigaJoule (GJ, unit of energy) in pre-industrial systems fall into a low and a high output subgroup, the output of the low subgroup, comprising 5 of the 8 systems, are one-twentieth to one-fifth of the full-industrial yield. *However, the output of the high subgroup are 2 to three times those of the full-industrial systems.* The yield of semi-industrial systems are more homogeneous, with an average of 51.75 GJ, while the yield of full-industrial systems, even more uniform, average 65.99GJ.

When the ratio of total energetic output to total input is examined, however, pre-industrial systems range between 7 to 10, with the figures for the most productive systems being as high as 15 to 28. Semi-industrial systems gave ratios of 2 to 9, whereas the ratios of full-industrial systems are not much better than unity. These figures illustrate the law of diminishing returns remarkably well: there seems to be a plateau of output per hectare around 70-80 GJ, which is only rarely exceeded, as in the 3 high yielding preindustrial systems of Yunnan, China. Intensification of energy input leads to a drop in efficiency which is particularly sharp as input approaches the output ceiling. This drop in efficiency reflects the increasing entropic costs of high rates of dissipation, which occurs when the rate of energy input exceeds the capacity of the system to store the energy, as mentioned in the last Section. The exceptionally high output of the Chinese systems is also an indication that the energy storing capacity of a system *can* increase, depending on the space-time differentiation and the dynamic closures introduced. For example, the utilization of farmyard and human manure as organic fertilizers, which has been traditionally practiced in China, will have the effect of increasing dynamic closure (see also Section 3.5).

The living system depends, not just on stored energy, but stored *mobilizable* energy. Hence energy stashed away that cannot be mobilized is unavailable for work. As such, it is also equivalent to entropy. Thus, a country like Britain, dominated by the financial markets, and 'rentiers' who expect high returns from investments and underinvest (Hutton 1995), bears all the hallmarks of a system with a high entropic load. Much of Britain's investment is also diverted overseas to exploit cheap labour costs or cheap resources, encouraging environmental destruction in the Third World. This further increases the entropic load on the global ecological system, which, as explained above, feeds back on the world economic system to make it poorer on the whole.

4. Symmetrical Coupling and Space-Time Structure in Economic Systems

Symmetrical coupling of energy mobilization is a hall-mark of the healthy organism. It is that which enables the system to mobilize energy at will and in a perfectly coordinated way. It involves certain reversibility of flows and reciprocity in relationships. What would these be for the economic system? It would be a relationship of trust and goodwill, of cooperation, of connectivity in the system, so that intercommunication is optimized. It would be a smooth and balanced coupling of production to consumption, of employer to employee, of lending to borrowing, of investment to modest profit-taking. It is based on a differentiated space-time structure, and a respect for that structure, so that debts and surpluses can be properly distributed and re-distributed to offset one another and to maintain the system as a whole.

A slight digression into the living system will illustrate what I mean. For our muscles to work, even under the extreme exertion of say, long distance running, or better yet, a man running away from a tiger, energy has to be efficiently mobilized over a range of space-time scales. The immediate energy supply is in the form of the universal energy intermediate, ATP, which biochemists themselves have likened to "energy currency". *The important thing in a healthy system is that the ATP is never allowed to become depleted in the working muscle* (otherwise the man will be mauled and killed by the tiger). How is that accomplished? It is accomplished by a cascade of energy 'indebtedness' to more and more distant, longer term energy stores, which are replenished *after* the crisis is over, and the man can recover and have a hearty meal (see Ho 1995a). Thus, for the economic system to function effectively and efficiently, it has to achieve a smooth distribution and redistribution of surplus and indebtedness in space and time as the need arises.

In the free-market model, each player is supposed to out-compete every other, in order to reap the maximum benefit in the shortest time. The assumption is that the players on both sides of the relationship are equal so that if the producer produces shoddy goods and charge high prices, consumers will shop around and only buy the best and cheapest available, and so an equilibrium will eventually be reached. Even if that were true, a lot of unsustainable wastage or dissipation would have been generated before the equilibrium is reached. More to the point, real world economies consist of big, multinational companies as producers/employers who see it as their job to reap maximum profit for their shareholders, who, in any case, do not have any say in running the companies; and powerful banks as lenders who may charge high interest rates and expect quick repayment. In the absence of proper regulation, as in Britain's deregulated economy (see Hutton 1995), those are just the conditions for uncoupling and dissipation.

Hutton has contrasted Britain with a number of different economies including Japan and Germany, which are deemed to be more successful than Britain. It is of interest that both Japan and Germany go to great lengths to create and foster cooperation and trust, between employers and employees, among different companies (through cross share-holdings in Japan) and between companies and banks, which make much more favourable, long term loans to companies, respecting the realistic timescales required for companies to mature.

In Germany, employees are represented on the board of directors of companies and participate fully in making decisions. This has meant that Germany has largely avoided the uncoupling effects of long-term strike actions. Britain, by contrast, has had an adversarial employer-employee relationship which led to the growth of trade union powers in post-war years. The Thatcher era succeeded in breaking the unions only to create the greatest uncoupling of all, the high unemployment rate, with irreplaceable talents and skills going to waste, not to mention the burden of human suffering and social alienation. Unemployment and poverty threatens the survival of the system as a whole.

5. Dynamic Closure and Sustainability

As emphasized earlier, a thriving economic system has two branches, the cyclic, non-dissipative branch, which requires dynamic closure, and the irreversible, dissipative branch. Dynamic closure is the key to a self-reproducing or sustainable economy. This principle has been well-appreciated by traditional indigenous farmers in their "internal input farming system" (Shiva 1991), which essentially involves closing the nutrient cycles of plants and animals as much as possible. It depends on a reciprocal, symbiotic relationship between farmers who farm and tend, and propagate the animals and plants, which in return provide sustenance for them and their community. It involves the minimization of waste by judicious recycling of nutrients, and diversification in the utilization of resources. By contrast, the so-called "high yielding varieties" introduced by the Green Revolution are designed to break nutrient cycles, dispense with recycling to depend on intensified external input. The lack of dynamic closure and the intensification of external input are the reasons why fully industrialized agricultural practices are so energy inefficient, and ultimately nonsustainable.

Dynamic closure depends on cycles of human reproduction, reproduction of life-stock and crop-plants, manufactured goods and so on, integrating with cycles of investment and reinvestment for self-renewal and maintenance, if not for growth. The cycles are interlocked, or catenated. Present generations work, not just for themselves, but for their children's and parent's generations. Past debts are being repaid, or surpluses consumed, while present debts or surpluses are being created for the future. These circulation of resources in space and time are essential for the survival of the system as a whole. Each generation invest and re-invests towards the immediate and distant futures.

When these cycles are broken, the dissipative branch inevitably increases at the expense of the non-dissipative, and threatens the sustainable reproduction of the system itself. Re-investment must occur in the business/industrial sector, as much as in housing, public transport and most of all, in education.

Skilled labour, as well as professionals and managers have to be trained and retrained to replace the aging or obsolete generations. UK industry has had a poor record in investing in people, despite the recent initiative by the very name. The depletion of "human capital" means that the system effectively wastes away. There is indeed a desperate need for renewing and reconnecting the life-giving, life-sustaining cycles.

IV. Conclusion

I have outlined a theory of the living system as a dynamically closed, self-sufficient energetic domain of cyclic non-dissipative processes coupled to irreversible dissipative processes. *Mobilizable* energy is stored over all space-time domains, so that intercommunication is optimized, enabling the system to function as a coherent whole.

I show how this conceptual framework can provide diagnostic criteria for a healthy organism and by extension, a healthy economic system. A healthy organism maximizes cyclic, non-dissipative, symmetrically coupled flows while minimizing irreversible, dissipative flows of energy and resources. It maximizes space-time differentiation, and dynamic closure so as to increase energy storage within the system. This leads to new insights concerning the equivalence of money to *both* energy and entropy, such that economies cannot be tuned simply by controlling the money supply. It also makes explicit the role of trust, cooperation and goodwill in fostering symmetrically coupled flows, and the importance of re-investment in all sectors in achieving dynamic closure of the system on which a self-reproducing, sustainable economy is ultimately dependent.

References

BARBER, W.J.: *A History of Economic Though,*, London (Penguin) 1967.

BECKER, G.S.: *Human Capital: a Theoretical and Empirical analysis with Special Reference to Education*, New York (NBER) 1966.

BREITHAUPT, H.: "Biological Rhythms and Communications", in: F.A. POPP, U. WARNKE, H.L. KONIG, W. PESCHKA (Eds.): *Electromagnetic Bioinformation 2nd ed.*, Berlin (Urban & Schwarzenberg) 1989, pp. 18-41.

CROWTHER, G.: *An Outline of Money*, London (Thomas Nelson and Sons Ltd) 1940.

DEANGELIS, D.L.: *Dynamics of Nutrient cycling and Food Webs*, London (Chapman and Hall) 1992.

GHISELIN, M.: *The Economy of Nature and the Evolution of Sex*, Berkeley (University of California Press) 1974.

HAKEN, H.: *Synergetics*, Berlin (Springer-Verlag) 1977.

HAROLD, F.M.: *The Vital Force: A Study of Bioenergetics*, New York/San Francisco (W.H. Freeman) 1986.

HO, M.W.: *The Rainbow and the Worm: The Physics of Organisms*, Singapore (World Scientific) 1993.

HO, M.W. (1994a): "What is Schrödinger's Negentropy?", *Modern Trends in BioThermoKinetics*, 3 (1994), pp. 50-61.

HO, M.W. (1994b): "Towards An Indigenous Western Science - the Organism as a Coherent Space-time Structure, in: W. HARMAN, J. CLARK (Eds.): *New Metaphysical Foundations of Modern Science*, Sausalito (Institute of Noetic Sciences) 1994, pp. 179-213.

HO, M.W. (Ed.) (1995a): *Bioenergetics, S327 Living Processes, An Open University Third Level Science Course,* Milton Keynes (Open University Press) 1995.

HO, M.W. (1995b): "Bioenergetics and the Coherence of Organisms", *Neural Network World*, 5 (1995), pp. 733-750.

HO, M.W. (1996a): "Evolution", in: G. GREENBERG, M. HARAWAY (Eds.): *Encyclopedia of Comparative Psychology*, New York (Garland Publishing), 1996 (in press).

HO, M.W. (1996b): "The Biology of Free Will", *Journal of Consciousness Studies,* 1996 (in press).

HO, M.W. (1996c): "Bioenergetics and Biocommunication", in: R. PATON (Ed.): *Proceedings IPCAT95*, Singapore (World Scientific), 1996 (in press).

HO, M.W. (1996d): "Towards a Theory of the Organism", 1996 (submitted).

HO, M.W. (1996e): "Natural Being and Coherent Society", in P. BUNYARD (Ed.): *Gaia in Action - Science of the Living Earth*, Edinburgh (Floris Books) 1996, pp. 286-307.

HO, M.W., S.W. FOX (Eds.): *Evolutionary Processes and Metaphors* , London/New York (Wiley) 1988.

HO, M.W., SAUNDERS, P.T. (Ed.): *Beyond Neo-Darwinism: Introduction to the New Evolutionary Paradigm*, London/New York (Academic Press) 1984.

HODGSON, G.M.: *Economics and Evolution: Bringing Back Life into Exonomics,* Cambridge (Polity Press) 1993.

HUTTON, W.: *The State We're In,* London (Vintage) 1995.

KOSLOWSKI, P.: *Ethics of Capitalism and Critique of Sociobiology*, Berlin (Springer) 1996.

KUMP, L.: "Geophysiological Society: Mission Statement", 1996.

LOVELOCK, J.: *A New Look at Gaia,* Oxford (Oxford University Press) 1979.

LOVELOCK, J.: "The Gaia Hypothesis", in: P. BUNYARD (Ed.): *Gaia in Action, Science of the Living Earth*, Edinburgh (Floris Books) 1996, pp. 15-33.

MAYNTZ, R. "The Influence of Natural Science Theories on Contemporary Social Science", in: M. DIERKES, B. BIERVERT (Eds.): *European Social Science in Transition*, Boulder Colorado (Westview Press) 1992, pp. 27-79.

MOROWITZ, H.J.: *Energy Flow in Biology*, New York/London (Academic Press) 1968.

ORMEROD, P.: *The Death of Economics*, London/Boston (Faber & Faber) 1994.

PIMM, S.L.: *Balance of Nature - Ecological Issues in the Convservation of Species and Communities*, Chicago (The University of Chicago Press) 1991.

PRIGOGINE, I.: *Introduction to Thermodynamics of Irreversible Processes*, New York/London (John Wiley & Sons) 1967.

SAUNDERS, P.T.: "Evolution without Natural Selection: Further Implications of the Daisyworld Parable", *Journal of theoretical Biology,* 166 (1994), pp. 365-373.

SHIVA, V.: *The Violence of the Green Revolution*, Penang, Malaysia (Third World Network) 1991.

SCHNEIDER, E.D., J.J. KAY: "Life as a Manifestation of the Second Law of Thermodynamics", *Mathematical and Computer Modelling*, 19 (1994), No. 6-8, 25-48.

SCHNEIDER, E.D., J.J. KAY, M.W. HO (1996) (in preparation).

SCHRÖDINGER, E.: *What is Life?* Cambridge, Mass (Cambridge University Press) 1944.

WHITEHEAD, A. N.: *Science and the Modern World*, Harmondswoth (Penguin) 1925.

WITT, U.: "Self-organization and Economics - What is New?", preprint, 1996.

ZENG, H.K., Z.W. QIAN, M.P. MYERS, M. ROSBASH: "A Light-entrainment Mechanism for the *Drosophila* Circadian Clock", *Nature*, 380 (1996), pp. 129-135.

Chapter 12

Darwinism and Economic Theory*

PETER T. SAUNDERS

I. Introduction

There has always been a close connection between evolution theory and economics. Both Darwin and Wallace acknowledged their debt to Malthus, whose famous observation that populations have the potential to increase faster than the resources on which they depend became one of the axioms from which the theory of natural selection is derived. The influence of economics did not stop there, of course, but continues to the present day. Terms such as cost-benefit analysis and investment are in common use in evolution, and with their usual meanings. One evolutionist has even described his own book as a "cross between the *Kama Sutra* and the *Wealth of Nations*" (Ghiselin 1974, p. 1). Conversely, many economists, for example Marshall, Veblen and, more recently, Friedman (1953), have looked to Darwinism for ideas.

* I am grateful to Mae Wan Ho and Ulrich Witt for comments on an earlier draft of this paper.

A particularly striking example is provided by game theory. While this was was originally developed for economics by von Neumann and Morgenstern (1944), much of the economists' recent interest in the subject arises out of its application to evolution theory (Maynard Smith/Price 1973). Even where the links are not quite so obvious, there can be no doubt that the two subjects have profoundly influenced each other. Indeed, given the sorts of subject matter they address, it would be surprising if they had not.

There are at present debates going on in both evolution theory and economics, and here again there are many similarities. The aim of this paper is to explore these, starting from the situation in biology and noting the parallels in economics[1].

II. The Orthodox Theories

It would be easier to describe and compare the debates if each were between two well defined positions, but this is not the case. This is more obvious in economics, in which there are a number of different schools of thought located within the mainstream or orthodox theory. They are sometimes collectively referred to as neo-classical economics, but there has been a great deal of development since the time of Walras, some of it in quite diverse directions.

The vast majority of evolutionists, in contrast, would describe themselves as neo-Darwinists. This does not simplify the issue as much as we might expect, however, because there is no agreed definition of neo-Darwinism, and indeed it is really more a paradigm than a theory.

The core of neo-Darwinism is, however, not difficult to identify. According to its supporters, it involves the assumption of the "overwhelming importance of natural selection", (Mayr 1982), and the idea that "mutation, whereby variation arises, and selection, whereby it is shaped into coherent adaptive form, is considered sufficient explanation of the evolutionary process" (Futuyma 1977). (By *neo*-Darwinism is meant the synthesis of Darwin's original theory with Mendelian genetics. An alternative term is the "synthetic theory".)

1 For an economist's account of the connections between the theories, though not the current debate, see HODGSON (1993).

The key point is that the direction of evolution is determined not by the organisms (or species) themselves, nor even by the workings of the ordinary laws of physics and chemistry within them, but by the external forces that act on them, *i.e.* the selection pressures. It is these that hold the key to understanding evolution, rather than the organisms themselves.

This is very much like the picture in Newtonian mechanics. To understand why a planet goes around the sun in an approximately elliptical orbit we do not have to know much about the planet itself. The path is determined by an external force, the gravitational attraction of the sun[2]. What is more, we can make a number of gross simplifications without losing the essence of the problem. We can suppose that the sun and the planets are solid, static and spherical, which they are not, that the planets do not influence each other, which they do, and so on, and while the predicted trajectory will not be exactly what is observed, it will be close. What is more, there are standard techniques for incorporating the factors we neglected as minor adjustments to the model, and hence, by a process of successive approximations, making the predictions as accurate as we like.

Neo-Darwinists believe that evolution can be explained in much the same way. Natural selection can account for almost everything, certainly for all the important features. The role of any other factors (such as 'developmental constraints') is to explain away minor anomalies. The conventional view in economics is similar: while there may be other factors to take into account, the market is the starting point and the focus for explanation.

III. The Alternative in Biology

Many of the critics of both neo-Darwinism and orthodox economics start from what they consider the inherent implausibility of the claim that such simple theories can be adequate to explain such highly complex phenomena. It follows that the alternatives they propose cannot be described in a few short sentences. There is, all the same, more to the critics' case than mere

2 In general relativity, the situation is quite different. In particular there is no such thing as the 'force of gravity'. Biologists and economists who are attracted by the paradigm of physics should study the physics of the twentieth century, not the nineteenth. See SAUNDERS (1989), also HO (1993).

opposition to the present theories, and a brief sketch of one part of what is happening in evolution will serve to illustrate both that alternatives exist and also that they are not mere philosophical quibbles. They can and do lead to research which is capable of explaining important phenomena in a way that is different and, the critics would argue, more satisfactory than is possible within neo-Darwinism.

1. The Importance of Development

In evolution, the alternative is based on two fundamental premises. First, the organisms we see about us today are the products of a process of descent with modification, *i.e.* evolution, and this has occurred by the action of the known laws of science. There are no otherwise unobserved fields or forces. This is in complete agreement with the synthetic theory, and in opposition to Sheldrake's (1981) morphic resonance hypothesis, and, for that matter, creationism. Second, and this is where the difference lies, while natural selection does occur and does play a role in evolution, this is more as editor than as creator. Darwin was not wrong, but his theory, even in its modern form, is not adequate to account for evolution.

This is not a new idea; it has been around for almost as long as Darwinism itself. It was well expressed by D'Arcy Thompson (1917), who wrote in his famous book *On Growth and Form*:

"The waves of the sea, the little ripples on the shore, the sweeping curve of the sandy bay between the headlands, the outline of the hills, the shape of the clouds, all these are so many riddles of form, so many problems of morphology, and all of them the physicist can more or less easily read and adequately solve: solving them by reference to their antecedent phenomena, and in the material system of mechanical forces to which they belong and to which we interpret them as being due....

"Nor is it otherwise with the material forms of living things. Cell and tissue, shell and bone, leaf and flower, are so many portions of matter, and it is in obedience to the laws of physics that their particles have been moved, moulded and conformed."

Darwinists accept that physics and chemistry are involved in bringing organisms into being. They can hardly deny that. They do not, however, see them as relevant to explaining why they have the forms they do. For them, developmental biology is the study of construction, not of architecture and design. And this is one of the key points of disagreement between neo-Darwinism and the new approach.

To see what this means in practice, consider how we might explain the shapes of the medusoids illustrated in Fig. 1. A neo-Darwinist account would be concerned with showing that a medusoid gains – or might gain – a selective advantage from such a shape. It might also include a story of how the medusoid could have evolved from a simpler ancestor through a sequence of intermediates, each slightly more complicated and slightly fitter than the one before.

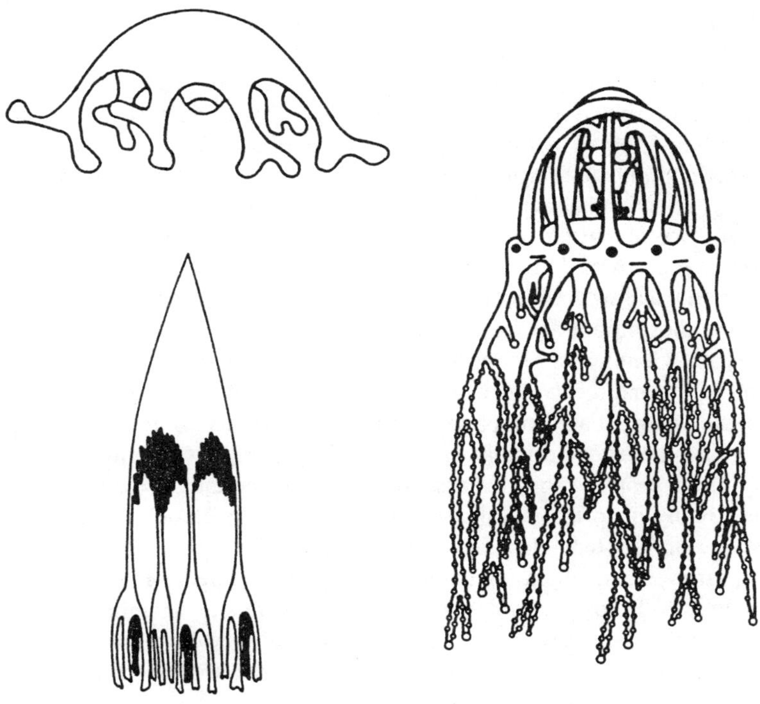

1. *Cladonema*, *Cordylophora*, and a drop of fusel oil in paraffin. Redrawn from D'Arcy Thompson (1917).

Actually, only two of these are drawings of medusoids. The one on the lower left is not a medusoid. In fact, it isn't an organism at all. It is a drop of fusel oil which has been allowed to fall into paraffin. Now a drop of fusel oil doesn't have ancestors, it doesn't have a genome, and it isn't subject to natural selection. Its shape is determined solely by the physical forces that act on it as it enters the paraffin.

Once you know that, does it still seem that the most important part of the explanation of the shape of the medusoid lies in selective advantage? Surely not. The undirected physical forces between two fluids of appropriate densities, viscosities and surface tensions, and with the right relative velocity, can produce the shape. It is therefore one that can appear relatively easily and reliably in nature, even though that isn't obvious at first sight.

It may well be that this peculiar shape confers some advantage on the medusoid. Certainly if it were a severe handicap it would have been eliminated by natural selection, but this explains only why it has persisted. It tells us neither how it arose in the first place nor why the medusoid doesn't have some other complicated form instead of this one. Are we really supposed to believe that selection has acted on a long sequence of small random variations only to produce one of the relatively small number of forms that are also observed in inert drops of fluid?

Selection was involved in the evolution of the shape of the medusoid, but it is not the most important factor in understanding why medusoids look the way they do. Genetic variation may be effectively random but phenotypic variation is not, and it can play a directing role. (See Saunders 1984)

The explanation based on physics has more to tell us as well. The shape of an oil drop depends critically on the conditions of the experiment. If the velocity of the drop as it hits the paraffin is below a certain threshold, it will become slightly distorted, but that is all. Above that threshold, there appear not just one or two 'tentacles', but many. This sort of behaviour is common in nonlinear systems, which typically possess one or more so-called bifurcation parameters. Below a certain value of such a parameter, nothing interesting happens; if it increases above that value, a completely different form may be produced. Like the straw that broke the camel's back, the crucial change in the bifurcation parameter need not be dramatic in itself and may be brought about by something very simple, such as slightly increasing the height from which the fusel oil is dropped.

There is also usually more than one way of affecting a bifurcation parameter. In this case, we could also change the temperature of the oil or the paraffin to alter their physical properties. The important point is that it doesn't matter how we change the bifurcation parameter nor exactly how great the change is. If the parameter does not cross the threshold, nothing much will happen. If it does, then there will be a significant change which will be almost totally independent both of the precise size of the change in the parameter and of what brought that change about.

This has many important consequences. One of them is that for the same phenotypic change to occur in different individuals does not require a change in the same gene in each, nor even in the same gene product. This can help to overcome the problem that a mutation that occurs in a single individual is very likely to be eliminated by genetic drift even if it confers a significant selective advantage. It is improbable that the same random mutation will occur in a number of individuals, but if many different mutations will have the same effect, *i.e.* if the variation is a property of the developmental system which all individuals within a species share instead of a particular mutation which is unlikely to occur in more than one of them, then it is much easier to imagine that it will persist.

The bifurcation parameter can also be affected by changes in the environment, which accounts for the frequent occurrence of so-called phenocopies: individuals which, even thought they are genetically normal nevertheless resemble known mutants. This can help to solve the problem of how to explain evolutionary change when the (hypothetical) intermediate stages do not appear to be advantageous even though the end result is. A well known example, pointed out many years ago by Mivart (1871), is that flat fish that swim near the bottom of the sea have both eyes on the same side of the head. If evolution is always gradual, as neo-Darwinists insist it is, one of the eyes must have moved to its present position by a long sequence of small steps, each an improvement on the one before. It is very difficult to imagine what advantage there could have been in having one eye only slightly further forward but still pointing downwards. Yet if there were not, the transition to the modern form could never have begun[3].

To see how we can solve this puzzle, consider the simple case in which a useful variation requires two mutations. Each of them increases the bifurca-

3 In the sixth edition of the *Origin of Species*, Darwin acknowledged this point and suggested that the movement of the eye occurred because successive generations of fish strained their lower eyes to look upwards. His modern successors would dismiss such an account as Lamarckian, but they have no better explanation.

tion parameter by approximately half the amount required to take it over the critical value. If either mutation occurs alone, there will be no obvious phenotypic change, and apparently no reason for the allele to spread through the population. It would therefore seem to be only by extreme good fortune that any individual will possess both of the necessary alleles. The variation would be selected if it were to occur; the problem is that it apppears very unlikely ever to occur and be available for selection.

On the other hand, the presence of either mutation reduces the change in the bifurcation parameter required to bring about the variation. It is therefore more likely to occur as a phenocopy. In this way, an individual with only one of the mutations can have a selective advantage. Even if part of a variation is of no benefit, an increased probability of having the whole variation can be. The second mutation, when it occurs, would lock in the variation, which would then no longer require the environmental stimulus: this is the process known as genetic assimilation (Waddington 1940).

When a bifurcation parameter is increased slowly, typically very little happens for a while, then there is a dramatic change, and after that nothing much happens again. This is just the pattern which many palaeontologists claim is observed in the fossil record and which has been given the name 'punctuated equilibria' (Eldredge/Gould 1972). There has been a considerable controversy over this in evolution theory because one of the basic tenets of Darwinism is that evolution has to be gradual, not because of the evidence but because the theory predicts there will be no sudden changes. The alternative not only allows them, it actually suggests that the major changes in evolution are likely to have involved some large steps (Saunders 1993).

2. Mimicry

The Viceroy and Monarch butterflies are both foul tasting, so birds quickly learn to avoid them. They are not close relatives, but they look very much alike, and this gives them a distinct advantage. Because the birds have only one pattern to recognize, fewer butterflies of each species are taken while they are learning.

Any variant form of either species that did not have the characteristic markings would not be identified as foul tasting even by experienced predators until it was too late. Unless it had some other compensating advantage, this would make it less likely to survive and leave offspring. Thus natural selection can explain why the similarity persists.

It does not, however, explain how it arose in the first place, which is the more important question for a theory of evolution. It seems very odd that the Viceroy should have gone to such trouble to copy the complicated wing pattern of the Monarch. We can see the advantage once the transition is complete, but how and why did the Viceroy ever start down that path? Its ancestors are generally accepted to have had rather drab markings, as some of its relatives do today. It would surely have been far more effective to have evolved almost any conspicuous coloration at all, than to have changed to looking only slightly more like a Monarch.

Nijhout (1978) has shown, however, that the way that butterfly wings develop makes the transition from the pattern of the Viceroy's typical relatives (and hence, we suppose, of its ancestors) to that of the Monarch surprisingly easy, requiring only one or two changes in parameters and so, presumably, only one or two mutations. This makes it much easier to understand how the similarity arose in the first place.

It is not always the case that species that resemble each other gain an advantage from this. *Lexias aeropus* and *Anetia cubana* are two species of butterfly that are not even in the same family, yet the similarity between them is just as close as that between the Monarch and the Viceroy. What is more, they live on opposite sides of the Earth, the former in Indonesia and the latter in Cuba, which makes it highly unlikely that natural selection could have been responsible for the similarity; no birds would ever encounter both species.

Again, the explanation almost certainly lies in development. If most wing patterns are variations on a single theme (the so-called nymphalid ground plan) then it is not surprising if the same one turns up more than once, especially in similar environments. It is interesting that neo-Darwinists refer to this sort of resemblence as pseudo-mimicry, the implication being that any phenomenon they cannot explain is not really a phenomenon at all. It is of course very convenient for a theory if it can avoid awkward cases by redefining them out of existence.

3. Rational Taxonomy

One of the tasks of biology is to classify living things into a meaningful order. One obvious way of doing this is by relatedness, and indeed this is what most modern systematists set out to do. They use a number of techniques in their work, and there are strong disagreements among them, but the chief goal is to determine the true phylogeny.

As Ho (1990) argues, however, this begs the question of whether the genealogical classification is necessarily the most natural. Taxonomy arose in the first place because the forms of organisms seem to be arranged in orderly hierarchies of groups. But a taxonomy of forms is not the same as a phylogeny and cannot ultimately be based on comparisons of DNA sequences. It must be based on the forms themselves, and it can only be derived from experimental and theoretical studies of development, the process by which the forms are generated (Ho 1990; Ho/Saunders 1994).

The aim of such a taxonomy is to establish a map giving the "transformational distance" between any two forms, rather than a Hamming or other suitably defined distance between genomes. This allows us to make predictions about evolution, because change can be expected to occur between forms that are nearest neighbours on the taxonomic map, rather than predominantly between those that are connected by (possibly long) sequences of individually advantageous mutations[4].

There is thus a choice to be made. On the one hand, we can base our taxonomy on DNA and define evolution to be permanent change in gene frequency. Alternatively, we can base our taxonomy on form, which makes evolution effectively the study of phenotypic change. The former is the approach of most orthodox evolutionists (with the notable exception of Ernst Mayr); the latter is what most people outside the field imagine the study of evolution is about; the point is that they are not the same.

Thus the present controversy about evolution is not a matter of playing with words; it concerns real issues and implies a different approach to some of the most important questions the subject should be addressing.

---

4 This statement is equivalent to the principle of minimum increase in complexity (SAUNDERS/HO 1981).

IV. Some Parallels

Like evolutionists, economists place their emphasis on an external force, in their case the market. True, you need farms and factories and workers if you are actually going to produce anything, but there is no need to study them closely or put their details into the theory. We can assume that they will appear as and when they are called into play by the market. If there is a demand for widgets, entrepreneurs will set up widget factories and workers will become widget-makers. This is the same sort of confident assumption expressed by Dobzhansky *et al* (1977) in their textbook *Evolution*: "If the appropriate genetic variants to face an environmental challenge are not already present in the population, they are likely to arise soon by mutation."

Not all biologists agree with this, and in economics we find, for example, that Simon (1991) argues that many of the properties of modern capitalism have more to do with the structure and internal relations of non-market organizations than with markets.

In particular, some economists maintain that the future of an economy depends crucially on the presence of an educated workforce and an an appropriate infrastructure, and not just on the classical economic factors. The conventional idea is that the output of an economy depends on the amount of capital and labour employed. The theory also says that if you increase them both you should increase the output proportionately, but if you increase only one of them there is a law of diminishing returns.

It follows that poor countries should find it easier to grow faster than rich ones, so a plot of growth against current output should be a line with negative slope. This works for the Asian dragons and the rich countries, but not for the rest of the world. Underdeveloped countries are unable to respond as the model predicts they should. It's not that the market doesn't operate nor even that the orthodox theory of growth is without foundation. The problem is that there is a crucial factor being overlooked. This is a serious matter, because it has practical implications. A policy that might be appopriate in an advanced economy may be disastrous in a poor country that lacks the infrastructure necessary to make it work.

As another example, in the UK during the early 80's there was a sharp reduction in industrial capacity. The idea was to make industry more productive by forcing the less efficient firms into bankruptcy. Up to a point, this strategy succeeded, although many perfectly good firms were destroyed as well.

Unfortunately, in economics as in evolution, "survival of the fittest" is a tautology. The only way we can tell which are the fittest is by seeing which ones actually survive. These need not be the ones with some specific desirable property that we can identify in advance and that we would like to see persist.

Then, in the mid 80's, the British government sharply reduced taxes. According to the theory, there should have been a massive investment in factories to produce all the goods that people were now able to buy. Only that didn't happen. Instead, there was a boom in imports, which led to a huge balance of payments deficit. With so many factories closed and their work forces dispersed, the economy was simply not capable of responding to the market, certainly not fast enough to compete with overseas suppliers[5]. Modern national economies are, of course, open systems, and, as in thermodynamics (and other fields as well), this can lead to phenomena which appear counterintuitive if one's intuition has been built up by analysing closed systems. As Lambert *et al* (1986) point out, Kettlewell's (1973) classic study of industrial melanism in the peppered moth also ignored the significance of movement across the boundaries of the region being studied.

One point that will immediately strike a chord with an evolutionist is the claim made by many critics of orthodox economics (*e.g.* Ormerod 1994) that while economic theory may have something to say at the detailed level – the micro-level – its ability to understand the workings of the economy at the overall level – the macro-level – is "manifestly weak".

This is an exact parallel with the debate about macroevolution. No one doubts that the conventional neo-Darwinist theory can be used to explain at least some features of evolution. The question is whether the theory is capable of explaining the major changes. This is why there has been so much controversy surrounding the theory of punctuated equilibria, when you would have thought that only palaeontologists could get so worked up about whether the fossil record shows evolution happened gradually or in bursts.

There is another similarity which is striking, even if it isn't strictly in economics. The most famous of Margaret Thatcher's sayings was her claim that "There is no such thing as society." There are, according to her, only individuals and families. This is exactly parallel with the Darwinian view that

5 At least one economist, J.K. Galbraith, predicted what would happen, but more on grounds of common sense and the earlier experience of Chile than by using sophisticated theoretical analysis.

the motive force in evolution, selection, acts on individuals, not on groups – even to the point that through kin selection neo-Darwinism too considers the family as the largest unit that is relevant.

The market is also supposed to be the major force in employment. The idea is that an individual has a certain amount of time at his disposal, and he sells this to the highest bidder. This ensures that all jobs should get done and that the most demanding jobs should command premium rates of pay.

Some economists now argue that the idea of a labour market is at best a gross oversimplification[6]. People are not simply selling their leisure to obtain money to buy other things they value more. Jobs mean more than that; they have to do with status, with social relations, with self esteem. Being unemployed is not just failing to sell your labour, and its consequences for the individual are far more than a mere shortage of money. Nor is it true that as an employer you get what you pay for: in general you get either more or less, and which it is depends on non-economic factors such as morale and tradition.

In many countries there is a serious shortage of mathematics teachers, but this has not brought about a sharp rise in their salaries. As for the very high pay of company directors, does that result from the action of the market, or is it more that if some people are allowed to write their own pay cheques they will be tempted to write very large ones?

Labour is an example of a market that fails to clear, *i.e.* one in which prices do not adjust to the point where all the goods are exchanged. You can hardly claim that the labour market has cleared when unemployment rates are running at 10% or more. Once again, it's not that the law is totally wrong, even when applied to labour. The question that must be asked is in how many situations is it the major factor that determines what will happen, and is it enough to tell us how to manage an economy? An even better question would be "Is it enough to tell us how to manage a society?" so as not to prejudge the issue by assuming that economic factors are to be paramount.

6 See, for example, *The Labour Market as a Social Institution* (SOLOW 1990).

PETER T. SAUNDERS

V. Nonlinearity in Economics and Evolution

Over half a century ago, the embryologist C.H. Waddington (1940) was struck by the fact that developing embryos have a number of important and, on the face of it, surprising properties. Above all, the process is remarkably stable both against mutations and against environmental perturbations. The stability is of a special kind that biologists call homeorhesis: an embryo that has been perturbed slightly will usually not go back the state it was in at the time of the perturbation but will continue to develop and eventually reach the same state it would have attained had it been left alone. In other words, it is the developmental trajectory that is stable, rather than any individual state along it. If the perturbation is too large, the embryo may die, but it may also be transferred to a different, but also stable, trajectory. As a result, for many features the possibilities form a discrete set, not a continuum.

For example, butterflies that have certain mutations or have been raised at different temperatures fail to develop the normal wing pattern. The pattern that does appear, however, is not a random modification of the usual one, but one of a small number of known alternatives, which can be quite different from that usually observed.

The features that Waddington identified in developing embryos are characteristic of complex nonlinear systems as a whole. Since embryos are certainly members of that class, they are almost bound to have them. That's not to say that we should cease to wonder that there are such things as organisms, only that once we know that there are, we should not be surprised that they have these properties. It would be astonishing if they did not. (See Saunders 1993)

Homeorhesis, for instance, is just another name for what a mathematician would call the stability of an orbit. The existence of alternative trajectories is also a common feature of nonlinear systems. (Recall that a linear algebraic equation has a single solution but a quadratic has two.)

It follows that we cannot extrapolate our understanding of a small part of a complex nonlinear system to the whole. Each of the separate trajectories behaves for the most part very much like a linear system. As long as we are following one of them, there is nothing to warn us that the others exist. Yet we can go very wrong if we do not realize that they do. We are like one of the blind men in the story, each of whom had his own limited view of what an elephant is like. That is one of the major reasons why micro- and macro- are different in both economics and evolution.

We are also freed from the fatalism that conservative economists so readily embrace, the belief that any interference in the free market can only make things worse and will ultimately prove futile anyway. The slogan "There Is No Alternative" does not apply to nonlinear systems. For example, many economists now maintain there is a choice to be made between becoming a low skill/low pay economy and a high skill/high pay economy. A country, at least a relatively advanced economy like those of the EU, can decide which of the alternatives it will adopt by the decisions it takes about education and infrastructure and industrial policy. The market will certainly influence what happens, but only within the parameters which the nation may have the power to set.

Nonlinearity often provides an alternative explanation for phenomena that previously seemed to require the action of some external force, whether God, natural selection or the market. Heating a shallow pan of water from below produces the characteristic Bénard convection cells, the Beloussov-Zhabotinsky chemical reaction can create the same complicated patterns that we observe during aggregation by the cellular slime mold *Dictyostelium discoideum*, and so on. We now know that complex effects do not require complex causes[7]. In mathematics, the Mandelbrot set, a figure of immense complexity, can be generated by a computer program barely a dozen lines long.

Nonlinear systems can exhibit self-regulation, one of the properties that is considered characteristic of life and so commonly believed to be achievable only by conscious design or natural selection. (Watson/Lovelock 1983; Saunders 1994). Their internal dynamics make them far less sensitive than linear systems to short term external influences, whether natural selection or the immediate demands of the market. Nonlinear systems often behave in ways that seem paradoxical to the naive theorist; think of how a gyroscope tends never to move in the direction in which you push it. But while most real systems are nonlinear, linear systems are easier to model, and so it is on linear systems that most of our intuition is based.

---

7 This is in contrast to the so-called Curie principle of classical physics, according to which if effects manifest a certain symmetry, then this symmetry will be reflected in the causes that give rise to them. See STEWART/GOLUBITSKY (1993).

VI. The Problem of Good

Humans are not just passive entities, subject to market forces and nothing else. We are influenced by many other things as well. In fact, a good case can be made that if everyone acted as the market model would suggest, society would collapse. Rockefeller may have been right to say that to get one splendid American Beauty rose we have to cut off all the other buds, but he forgot about the rest of the plant, on which that single surviving rose depends. Not only can we not all be Rockefeller, society depends on the fact that most of us aren't trying to be.

It is interesting that Adam Smith did not take the same view as those who today see themselves as his successors and have adopted him as their patron saint. He did not see economic forces as the only determinant even of how an economy operates. And if morality and ethical behaviour did not come out of his model, Smith did not consider that a reason either for denying they exist or for claiming that the world would run better if they didn't. On the contrary, in *The Theory of Moral Sentiments* he explicitly stated his belief in a natural human tendency to act in a way that is acceptable to society. It is interesting that modern research shows that people really do value fairness: whether they will accept a certain wage or not, for instance, may depend as much on whether they think it is fair rather as on market forces as such.

Similarly, if evolutionists were not convinced that the theory of natural selection can explain every significant aspect of organisms, they would not find it so hard to understand why humans do not always act selfishly. Christian theologians have had to cope with the problem of evil: how can there be evil in a world created by a God who is both good and omnipotent? Sociobiologists, on the other hand, have to cope with what we might call the problem of good: how can there be anything but selfishness in a world created by natural selection? E.O. Wilson (1975) himself describes altruism as the central theoretical problem of sociobiology.

Not only are the problems similar, so are the solutions that have been proposed. The natural theologian William Paley, one of the scholars who most influenced Darwin, was careful only to write of "cases of apparent evil" (presumably it would have been sacrilege to accuse the Creator of allowing real evil into the world) and he argued that they were due to "the thwartings and crossings of laws whose effects are for the most part beneficient" (Paley

1819, p. 436). For sociobiologists, good is also only apparent – Ghiselin (1974, p. 247) writes, "Scratch an 'altruist' and watch a 'hypocrite' bleed" – and is explained as the occasionally subtle workings of laws whose effects are for the most part selfish. (For a critique of sociobiology, see Saunders 1988.)

Evolution theory and equilibrium economics each provide a means of justifying the status quo as the inevitable outcome of the workings of an Invisible Hand or, which is much the same thing, the Blind Watchmaker. In economics it is the market that determines what will be produced and at what price; in evolution it is natural selection that has created organisms and made them what they are. The individual agent or organism acts in accordance with these pressures in much the same way that the Earth follows the orbit determined by the gravitational field of the Sun.

Such beliefs allow us to reduce our personal responsibility for the way the world is and indeed for our own actions. True, as philosophers are constantly reminding us, "is" and "ought" are logically distinct. That does not, however, mean that they are not connected, nor that the ideas do not, rightly or wrongly, get conflated. The company director may not claim that his astronomical salary is morally justifiable – or, then again, he may – but it is very convenient for him that the theory says that it is necessary for the effective operation of the economy. The large tax reductions in Britain in 1987 were appropriate according to some interpretations of orthodox economic theory, but 1987 was also a year in which there was a general election.

It is also convenient for those of us who are, by world standards, very successful and affluent, to believe that this is because of our superior genes, and not largely a result of the exceptionally favoured environment into which we had the good fortune to be born. A person who knows he is behaving badly can rationalize his actions with the thought that they are, after all, natural. And how critical can the rest of us be if what he is doing is "only human nature"?

The trouble with an optimization theory is that it can be read backwards. If systems strive to move to the optimal state, then surely we must be in the best of all possible states, or at least well on the way towards it. Whether in economics or in evolution, this is a very attractive idea for those who are doing very nicely out of the world as it is.

VII. Conclusion

With the model of Newtonian mechanics very much in mind, if only implicitly, both biologists and economists have sought overarching theories that promise to explain a wide range of highly complex phenomena from a few basic principles. It is not by accident that we read of Darwin as the Newton of the grass blade, or Marxism as the laws of motion of society. The alternatives that are being proposed take the view that no such theories exist. Now that we know that there are solid mathematical reasons for doubting we can ever have three week weather forecasts even with excellent mathematical models, massive computers and dense networks of weather stations, why should we expect to be able to treat organisms or economies like the solar system? And even that, we are told, can exhibit chaos.

It is hard to imagine anyone ever trying to explain evolution without invoking natural selection, though we may come to speak more of selective constraints on development than of developmental constraints on selection. Equally we cannot ignore the effects of the market. The present parlous state of the former Soviet Union demonstrates simultaneously what happens both if you take the market too seriously and if you try to ignore it. The question is how far these concepts can explain the most important phenomena in their respective subjects and, consequently, how far they should be used as the unique basis for deciding policies.

When a young actor first joins a classical company to play in Shakespeare, he is often given the role of a "spear carrier". This is a minor character, often with no lines to speak, who may be necessary for the action but who has no influence on the way the plot unfolds. What is happening in both economics and evolution is that some of these spear carriers are being brought to the centre of the stage. Of course, the play will look quite different when they are.

References

DOBZHANSKY, Th., AYALA, F.J., STEBBINS, G.L., VALENTINE, J.W.: *Evolution*, San Francisco (W.H. Freeman) 1977.

ELDREDGE, N., GOULD, S.J.: "Punctuated Equilibria: an Alternative to Phyletic Gradualism", in: T.J.M. SCHOPF (Ed.): *Models in Paleobiology*, San Francisco (Freeman Cooper) 1972, pp. 82-115.

FRIEDMAN, M.: "The Methodology of Positive Economics", in: M. FRIEDMAN (Ed.): *Essays in Positive Economics*, Chicago (Chicago University Press) 1953, pp. 3-43.

GHISELIN, M.: *The Economy of Nature and the Evolution of Sex*, Berkeley (University of California Press) 1974.

HO, M.W.: "An Exercise in Rational Taxonomy", *Journal of Theoretical Biology*, 147 (1990) pp. 43-57.

HO, M.W.: *The Rainbow and the Worm*, Singapore (World Scientific) 1993.

HO, M.W., SAUNDERS, P.T.: "Rational Taxonomy and the Natural System as Exemplified by Segmentation and Phyllotaxis", in: R.W. SCOTLAND, D.J. SIEBERT, D.M. WILLIAMS (Eds.): *Models in Phylogeny Reconstruction*, Systematics Association Special Volume No. 52, Oxford (Clarendon) 1994, pp. 113-124.

HODGSON, G.M.: *Economics and Evolution*, Cambridge (Polity) 1993.

KETTLEWELL, H.B.D.: *The Evolution of Melanism.* Oxford (Clarendon) 1973.

LAMBERT, D.M., MILLAR, C.D., HUGHES, A.J.: "On the Classic Case of Natural Selection", *Rivista di Biologia* , 79 (1986), pp. 11-49.

MAYNARD SMITH, J., PRICE, G.R.: "The Logic of Animal Conflict", *Nature*, 246 (1973), pp. 15-18.

MIVART, ST G. J.: *The Genesis of Species*, London 1871.

ORMEROD, P.: *The Death of Economics*, London (Faber and Faber) 1994.

PALEY, W.: *Natural Theology* (16th edition), London (F.C. and J. Rivington and others) 1819.

SAUNDERS, P.T.: "Development and Evolution", in: M.W. HO, P.T. SAUNDERS (Eds.): *Beyond Neo-Darwinism: An Introduction to the New Evolutionary Paradigm*, London (Academic Press) 1984, pp. 242-263.

SAUNDERS, P.T.: "Sociobiology – a House Built on Sand", in: M.W. HO, S.W. FOX (Eds.): *Evolutionary Processes and Metaphors*, Chichester (Wiley) 1988, pp. 275-294.

SAUNDERS, P.T.: "Mathematics, Structuralism and the Formal Cause in Biology", in: B.C. GOODWIN, G.C. WEBSTER, A. SIBATANI (Eds.): *Dynamic Structures in Biology*, Edinburgh (Edinburgh University Press) 1989, pp. 107-120.

SAUNDERS, P.T.: "The Organism as a Dynamical System", in: F. VARELA, W. STEIN (Eds.): *Thinking about Biology*, Reading (Addison Wesley) 1993 (= SFI Studies in the Sciences of Complexity, Lecture Notes Vol. III), pp. 41-63.

SAUNDERS, P.T.: "Evolution without Natural Selection: Further Implications of the Daisyworld Parable", *Journal of Theoretical Biology*, 166 (1994), pp. 365-373.

SAUNDERS, P.T., HO, M.W.: "On the Increase in Complexity in Evolution: II", *Journal of Theoretical Biology*, 90 (1981), pp. 515-530.

SHELDRAKE, R.: *A New Science of Life*, London (Paladin) 1981.

SIMON, H.A.: "Organizations and Markets", *Journal of Economic Perspectives*, 5 (2) (1991), pp. 25-44.

SOLOW, R.M.: *The Labour Market as a Social Institution*, Oxford (Blackwell) 1990.

STEWART, I., GOLUBITSKY, M.: *Fearful Symmetry*, Oxford (Blackwell) 1992.

THOMPSON, D'ARCY W.: *On Growth and Form*, Cambridge, Mass. (Cambridge University Press) 1917.

VON NEUMANN, J., MORGENSTERN, O.: *Theory of Games and Economic Behaviour*, Princeton (Princeton University Press) 1944.

WADDINGTON, C.H.: *Organizers and Genes*, Cambridge, Mass. (Cambridge University Press) 1940.

WATSON, A.J., LOVELOCK, J.: "Biological Homeostasis of the Global Environment: the Parable of Daisyworld", *Tellus,* 35B (1983), pp. 284-289.

WILSON, E.O.: *Sociobiology*, Cambridge, Mass. (Harvard University Press) 1975.

Chapter 13

Evolutionary Economics and Evolutionary Biology

ULRICH WITT

I. Introduction

Present theorizing in economics is mainly directed towards two goals. First, there is the functionalist endeavor at explaining, or better rationalizing, why certain kinds of behavior and institutions can be observed in the economy. Second, efforts are made to attribute, in a hypothetico-normative manner, certain features, such as efficiency, equity, or optimality, to the outcome of the economic process. Both strands of thought relate to a definition of economics as a discipline which deals with the problem of scarcity. As a result of "nature's parsimony", as Ricardo once put it, humans (and probably not only humans, see Ghiselin 1978) have to use scarce means to achieve given ends. An alternative to this latently normative orientation is to interprete economics as a theory of human social behavior in the context of what are usually considered economic activities, namely production, accumulation, distribution, exchange, and consumption. From this perspective, explaining why and how the historically observable forms of production, accumulation, distribution, exchange, and consumption change so significantly and why the changes have accelerated so much in the last few centuries is a natural goal for economic theorizing.

It is precisely this 237 sphere as undergoing continuous change which has recently induced increasing efforts to develop an evolutionary approach to economics. Some economists at the turn of the century influenced by the 'Darwinian revolution' in natural history and the sciences in the 19th century had already expressed an interest in such an approach.[1] However, their pleas did not have much influence on actual economic theorizing. Darwin's -- and, even more so, Spencer's --thought seem to have been inspired by the idea of extending the upper classes' liberal social philosophy of Victorian Britain to nature (Young 1988; Bowler 1989, 1995; Desmond/Moore 1991; Richards 1992). The inspiration guiding economic theorizing at the end of the 19th century was quite different. The proponents of the 'neoclassical revolution' in economic theory (Walras 1954, Edgeworth 1881, Pareto 1896) tried to establish economic theory in terms of concepts borrowed from classical mechanics (see Mirowski 1988). In this they were turning their backs on both the pre-Victorian (Smith 1759, 1776) and the Victorian (Mill 1848) versions of social philosophy. They had a "mechanics of utility and self-interest" in mind in which the state of rest, and not processes of change, figured most prominently. Focus was on the (market) equilibrium in which not only all individual plans fit together, but in which all the agents' utility reaches an extreme value, a maximum in this case, subject to the mutually imposed constraints. (This is the dual formulation of the concept of equilibrium in classical mechanics where the state of rest is associated with minimum free energy.) The basic concepts and the way in which they were presented came from Newtonian physics -- the approach which the 'Darwinian revolution' had just begun to challenge (Mayr 1991, Chapter V).

In view of these markedly different intellectual developments and in view of the fact that an evolutionary approach is now more seriously being used in economics, it may be asked how relevant the 'Darwinian revolution' is for economics. In economics attempts to actually make recourse to Darwinian concepts have so far been made in two different ways. One is the direct application of the theory of natural selection to human economic behavior. The other is a metaphorical use of Darwinian concepts in the construction of

1 See the often cited introduction to MARSHALL (1898/1938); see VEBLEN (1898) and SCHUMPETER (1934); a good survey is in HODGSON (1993); for a most recent restatement of such an interest see KHALIL (1992).

analogies. As is to be expected, the different approaches imply entirely different answers to the question just posed. The present paper briefly outlines these approaches and discusses possible parallels with evolutionary biology. Section II focusses on the attempts at applying the theory of natural selection directly to the economic domain. Section III turns to contributions in economics in which the theory of natural selection is used as a metaphor. In section IV the theoretical background of a more indirect interpretation is sketched. Section V highlights some of the significant historical features of long term economic development and argues that the explanatory sketch given previously is helpful in making sense of the observed features of economic evolution in terms of a broader Darwinian world view. Section VI offers some conclusions.

II. Does Natural Selection Operate on Economic Behavior?

Attempts to extend Darwinism directly to human economic behavior have been made in contributions covering several topics such as game theory (Hansson/Stuart 1990, Güth/Yaari 1992) and social philosophy (Hayek 1967, 1979, 1988). The game-theoretic contributions are motivated by the fact that humans are a result of genetic evolution and that human behavior, in particular all kinds of behavior conducive to, or instrumental in, enhancing reproductive success, has been subject to natural selection. Following the lead of sociobiology in explaining the emergence of "reciprocal altruism" (Trivers 1971), evolutionary game theory has been applied to social dilemmas, particularly the prisoners' dilemma, arising in economic interactions. Conditions are derived under which, in a suitably defined human population, agents who share 'altruistic' genes have a selective advantage. They behave cooperatively even where, with self-interested preferences, rationality forbids such behavior. The contributions provide a logically sound explanation for the compatibility of altruism and rational decision making within a certain population where otherwise they are irreconcilable. It is the irreconcilability that gives rise to the dilemma. Taken by itself, this application of evolutionary game theory to human behavior can be interpreted as referring to early stages of human phylogeny during which selection pressure was intense. Preferences for behaving altruistically towards certain persons may have then been genetically fixed and thus still be present today. The argument has something to it and will be discussed further in section IV in a more general form.

The argument does not imply that the process described by evolutionary game theory is still operating on human behavior today. In fact, such an interpretation would be much stronger and difficult to defend. A vast array of inherited traits determines the potential behavior of each individual in a population of genetically reproducing organisms of a certain species. Genetic variety in the gene pool of the population translates into a distribution of traits and thus into potential behavior.[2] The case in which differences in the inherited traits of the organisms imply differential reproductive success and induce a change in the distribution of traits in the gene pool of the next generation is a situation of selection pressure. In the original, Malthusian, spirit of Darwin's theory, selection pressure may be considered to be reflecting a situation of competition between organisms for reproductive success or for items that are conducive to that success.[3] In the absence of selection pressure -- in a situation of affluence -- behavior which contributes to genetic fitness is no longer selected for. Behavioral variety may then increase and include variants which have no adaptive value. Thus, in a state of affluence, the Darwinian approach to explaining behavior by reference to its contribution to inclusive fitness is much less significant than under conditions of selection pressure. Given the abundance which modern economies have been able to create, it can scarcely be doubted that human behavior can vary significantly without having a differential impact on reproductive success. In particular, economi-

---

2 On the ontogenetic level, environmental influences (e.g. nutrition conditions) which affect the growth of the phenotypic organism may differ in their impact on actual behavior and thus affect the the extent to which this behavior lives up to its genetically determined potential. The ontogenetic differences are, however, supposed to have a minor and non-systematic influence on reproductive performance, i.e. the number of living offspring who carry the gene pool of the next generation, compared to the influence genetic differences have.

3 As, for example, the size and quality of a habitat. It is worth noting that competition for scarce resources may exert its influence through an ecology effect. Competition for feeding opportunities at the level of the predator may induce selection pressure at the level of the prey, if the adaptive features of the prey population vary as in the case of the evolution of industrial melanism.

cally more or less efficient forms of behavior do not seem to correlate with generating more or fewer offspring, not even when a trade-off between quality and quantity in raising offspring is accounted for.[4]

In a modified form, a direct application of Darwinian thought also underlies the theory of societal evolution outlined by Hayek. [5] He assumes three levels of evolution. The first is that of genetic evolution in which primitive forms of social behavior, preferences, and attitudes which bring about order in social interactions have been fixed genetically during man's phylogeny. Second, there is the evolution of the products of human intelligence and knowledge. Freed from the finite existence of each individual brain by efficient forms of coding, storing, and transmitting information, human knowledge has expanded enormously. Today the extent to which it makes the mastery of nature possible is impressive. Third, and this is what Hayek considers the core of his approach, there is the level of cultural evolution operating "between instinct and reason" (Hayek 1988, ch. 1). Culture, in Hayek's interpretation, is neither genetically conditioned nor rationally designed. It is a tradition of learnt "rules of conduct" (Hayek 1967) whose role, he claims, is often not understood even by those who follow the rules. They are passed on through cultural transmission - a 'blind' or spontaneous process in the sense that it is not consciously planned or controlled.

Hayek argues that the question of which rules come into being, and how this happens, is a matter of historical accident. In contrast, systematic influences that result from a selection process decide which of the rules survive. He believes that the process shapes cultural evolution and operates on groups, where the groups are defined as sharing the same rules of conduct. Those groups which succeed in developing and passing on rules better suited to governing their social interactions are able to grow and to feed a larger number of people. Their relative superiority may enable them to conquer and/ or absorb less well equipped, competing groups and thus, unintentionally, propagate the superior sets of rules. A growing population requires increasing specialization and division of labor. This, in turn, means that the underlying rules of conduct and the order they induce are extending. The rules become more differentiated and abstract (and, for the single mind, more difficult to

4 For further discussions see WITT (1985) and HERMANN-PILLATH (1991).

5 For statements of various aspects of that theory cf. the epilogue in HAYEK (1979) in which he explicitly draws on older notions of group selection in sociobiology; see also HAYEK (1988).

understand). An "extended order" is seen as spontaneously emerging in this way over centuries. It is what has enabled modern societies to achieve a historically unique level of civilization and productivity. This order embodies an impersonal intelligence accumulated during the selection processes in the form of surviving impersonal rules of conduct. Its most important achievement, Hayek (1988, chap. 3) submits, is trade and the emergence of a system of markets.

Thus, Hayek's theory is based on the twin ideas of spontaneous order as a system of impersonal rules of conduct and evolution as a result of group selection. Taken together they amount to claiming that natural selection not only 'chooses' between competing species, it also 'chooses' between the competing groups of humans in which acquired cultural norms and rules are fixed. The problem with Hayek's grand view is that it offers no explanation of the crucial processes of social learning and the transmission of the rules of conduct. How do these processes operate? Is their continuity sufficient to ensure that group selection can indeed exert a differential effect on the replication of the rules? Since much of the process is tacit, is it possible that "observational" or "vicarious" learning (Bandura 1986), i.e. learning by imitation, plays a crucial role? How does this fit the standard interpretation of group selection? These questions do not deny the possibility that social or collective learning, which takes place within stable groups of agents, is an essential element of economic evolution and may yield a basis for deriving appropriate transition laws (for a corresponding modeling effort see, e.g., Witt 1996). The point is, rather, that these processes of transmitting knowledge within a human population are too little understood to justify the far-reaching conclusions Hayek wants to draw.[6]

6 The co-evolutionary approach suggested, among others, by BOYD and RICHERSON (1985) covers several ideas of Hayek's theory in detail and is quite explicit in deriving corresponding transition laws. However, where it is not based on analogy, its crucial assumption is that social learning processes and genetic evolution interact or interfere with one another. Implicitly, this view presupposes that the theory of natural selection can be applied, albeit in a biased form, to present day human behavior, an interpretation that has been rejected here for the reasons given above.

III. Natural Selection as a Metaphor

Attempts to construct analogies between genetic and economic evolution (e.g. in Boulding 1981; Hirshleifer 1982; Matthews 1984; Allen 1988; Metcalfe 1989; Faber/Proops 1990; Saviotti/Metcalfe 1991; Hodgson 1995; Anderson 1994; Gowdy 1994, chap. 5) seem to be more popular than attempts to apply the theory of natural selection directly, or so it seems from their frequency. Yet there are serious objections to this kind of transfer too. To demonstrate this, imagine, for the moment, that the metaphor is used for the economic context. What then would the objects of economic selection be, or what would have to be thought of as the evolving entity in the domain of economics? Quite diverse answers to this question have been suggested in the literature. Some studies, for instance, focus on human artefacts such as mechanical devices for power transmission (Usher 1954) or products which serve certain particular needs (Saviotti 1988). The history of such artefacts is often well-documented. The Comptoise clock, which was produced from 1680 to 1914 in a small part of Eastern France, is a telling example (see Schmitt 1983).

The basic mechanical layout of that pendulum clock has changed only in minor ways. The transition from manufacturing by hand (an estimated 20000 pieces in the roughly first hundred years) to mechanized mass production (an estimated 2 million pieces in the following hundred years) has been accompanied mainly by changes in the outer appearance of the clock and the materials used. There have been several generations of faces, pendulums, and devices for striking the hour. The evolution of this artefact can thus be seen in the changes in its production technology, its mechanical layout, and its appearance. Its 'population dynamics' were significant before it eventually became extinct. Nonetheless, it is hard to provide a coherent explanation of the development in terms of a selection model. Many of the changes may be due to limitations on the materials available or to manufacturers' costs. Other changes may have been introduced by the manufacturers in anticipation of changes in fashion, i.e. biased adoption decisions of the consumers. A general causal structure does not seem to underlie these changes nor is there any hint of a continuously operating selection mechanism. The development of other artefacts has had different features so that the chances of finding a coherent explanation on the basis of the selection metaphor appear negligible.[7]

Another answer to the question of what might be the object of economic selection, or the evolving entity in the domain of economics, has been given in the literature on firms' organizations and their growth, persistence, decline, and eventual exit from the market. These contributions grew out of the debate on "economic natural selection" which has been instrumental in the development of evolutionary economics.[8] The question raised was whether, and to what extent, diversity in the firms' goals and performances would be eliminated by competition. Would all those forms which are not sufficiently profitable be driven out of the market? The idea was explicitly derived by analogy with Darwinian thought. However, for logical reasons, the analogy is difficult to defend. First, no convincing analogue for the two separate, systematically interacting, levels of the genotype and the phenotype can be found in the economic domain, a point already made by Winter (1964). Second, there is no entity comparable to a gene pool in terms of continuity or, more specifically, no equivalent to the continuity of the germ plasma. As a consequence, the notions of selection and replication refer to much less constrained and structured processes than in the Darwinian example. In evolutionary economics they therefore lack much of the significance they have in evolutionary biology.

Despite these reservations, the weakened form of the analogy suggested by Nelson and Winter (1982) is, up to the present, probably the most influential contribution to evolutionary economics. They argue that firms are organizations which have to base their internal interactions on behavioral routines, rules of thumb, and regular interaction patterns. The firms' behavior in production planning, calculation, price setting, and even the allocation of R&D funds is viewed as rule bounded. The corresponding "routines" are interpreted as the 'genotypes' while the firm's specific decisions thus derived

7 Several aspects of the changes which occurred in this and many other artefacts may, however, be captured quite well by a diffusion approach not based on any notion of selection, see ROGERS (1983), METCALFE (1988), NAKICENOVIC/GRÜBLER (1991). As explained in more detail elsewhere (WITT 1992), the frequency dependency effect is a dominant feature of diffusion processes and represents an important element in formulating respective transition laws.

8 ALCHIAN (1950), PENROSE (1952), FRIEDMAN (1953); among the few empirical studies are WEDERVANG (1965) and, more recently, AUDRETSCH (1995).

are interpreted as 'phenotypes'. The latter may be more or less favorable for the firm's overall performance measured in terms of profitable growth. Assuming that routines which successfully contribute to growth will not be changed, the actual expansion can be understood as an increase in relative frequency of those 'routine-genes'. Routines which cause deterioration in the firm's performance, on the other hand, are unlikely to replicate. The development of behavioral routines is thus considered to be a significant instance of economic evolution and an object of selection processes. Yet, there is little to support the assumption that this development is governed by a continuously operating selection mechanism (perhaps even one following transition laws given by a time-dependent change of respective frequency distributions). On the contrary, it is most likely that firms facing deteriorations are induced to identify the deficient routines and to replace or improve them in a kind of intentionally produced mutation of their 'genes'.

Nelson and Winter (1982) argue that such improvement or replacement activities are themselves subject to (higher) routines, and that the differential growth argument therefore applies once again. It may indeed be that firms activate higher problem solving routines on a regular basis when a crisis occurs or is expected to occur. Yet, the outcome of these problem solving routines is often more likely to depend on whose problem solving capacity is involved in those routines than on the routines themselves. In this case, the differential success of the problem solving process triggered cannot be derived from differences in those routines. Differential success is basically unpredictable (much like mutation in the genetic sphere). When different people are involved in the same routines at different times, development may be different. The continuity of what has been suggested as the analogue of the selection mechanism is lost. The argument derived from the selection metaphor rests on shaky foundations.[9]

9 The conclusion which AUDRETSCH (1995, p. 186) draws from his empirical findings is worth noting in this context: "..the main message ... is that, with respect to the dynamic patterns of firms over time, there is, in fact, no tendency that can be generalized. Rather, the dynamic nature in which firms and industries tend to evolve over time varies substantially from industry to industry. And there is at least some evidence that it is differences in the knowledge condition and technology underlying the specific industry, that is the nature of innovative activity, that account for variations in industry evolution across markets".

In a more general perspective there is yet another reason for the limited usefulness of the selection metaphor in the economic domain. In the genetic context, selective forces operate on a given population and change the relative frequencies of the genes in the pool. The individual members of the population have little, if any, room for escaping from these pressures or for adjusting to them. In this sense, selection forces may be labeled 'external'. In a similar way, the anonymous competitive forces of the markets can be seen as external to the individual. The market participants unintentionally impose constraints on each other in their multilateral interactions, and these constraints may even force some of them out of business. Thus the population of agents on the supply and/or demand sides of a market may systematically change over time. Yet, unlike in the genetic case, the agents are not helpless when exposed to these changes. Indeed, an important part of the systematic changes observable in markets, both in the composition of the participants and in their activities, may be attributed to the individuals' attempts to anticipate the effects of the market forces and to take account of them. If successful, the effects of 'external' selection are replaced by what would have to be labeled 'internal' selection - an attempt to change activities deliberately and so to avoid unfavorable selection consequences.

Since this dimension of economic behavioral adaptation has no equivalent in the genetic domain, the selection metaphor may be a misleading heuristic for understanding the crucially important feature of economic evolution - the role played by learning and cognition. In fact, the individuals' capacity to anticipate consequences of their behavior and to take account of them deliberately by searching for better solutions is more akin to Lamarckian than to Darwinian evolution. This is the reason why economic activities are much more volatile. Nonetheless, adaptive features are present here as well and they may be conjectured to follow their own regularities. Because of the limited human information processing capacity, these regularities are likely to reflect mental, rather than genetic, selection processes -- people are forced to be selective in what they sense, learn, and perceive. Since a coherent theory of these mental processes is still lacking, the present understanding of 'internal' selection processes and problem solving leaves many questions open. In any case, it may be inferred from the insights already available that the anticipation of external selection forces by internal selection, i.e. cognitively guided behavioral adaptations, does not necessarily lead to the same outcome as external selection alone. Cognitive anticipation usually allows for a multiplicity of solutions to a problem. This is a forteriori true where the invention of new, as yet unknown, solutions is considered.

IV. Economic Preferences and the Genetic Inheritance[10]

Even though there does not seem to be any evidence that natural selection forces govern the rapid adaptations of economic behavior observable in modern societies they may have an indirect effect. During human phylogeny, natural selection did shape human genetic endowments which, because those selection forces are no longer a source of systematic change, are still present and thus may have an impact on economic behavior. Among these endowments the basic, apparently genetically coded, conditioning or learning mechanisms and their "primary reinforcers" (Skinner 1966) seem to be particularly relevant. Before explaining their influence on economic activities it is necessary to elaborate briefly on the notion of reinforcers developed in psychology and on the notion of preferences considered in economics as a determinant of human decision making.

In the economists' view, individuals are supposed to order, according to their subjective preferences, alternatives which are available for choice. When a choice has to be made, people decide in favor of the most preferred of their alternatives. Under certain conditions, preferences can be represented by a utility function in which the independent variables are quantities of the objects of choice. Depending on the particular choices available, any objects can, in principle, enter the utility function. Since the individuals' subjective preference orderings are not directly observable, laboratory revelation procedures must be designed in which the ordering on a set of objects (as given by the experimentor) can be inferred indirectly from the observed choices. Although an individual's preferences may be viewed as reflecting her/his desires, economic theory usually refrains from attempting to specify, in a more general way, the objects from which utility is derived. With reference to the subjective nature of their evaluations, the question of whether individuals have cer-

10 For a more detailed discussion of the material of this section cf. WITT (1991).

tain preferences for objects in common, which would be expected when preferences have a genetic basis, is left open.

It is interesting now that, in developing a theory of revealed preferences, economists have tried to design basically the same experiments which behavioral psychology has set up for a different purpose, namely when an attempt is made to empirically determine reinforcers, that is, those items which have the power to reinforce operant behavior of an organism (Lea 1983). If a preference revelation experiment were run with mammals, some basic items could therefore be expected to be found as objects of the animals' preferences - those which have been identified in laboratory experiments with mammals (Millenson 1967) as primary reinforcers: air, water, sleep, warmth, nutrition, sexual activity, maternal care, love and affection, physical activity, and novelty. Because these elements are held in common and because of their obvious adaptive value, they may be considered part of the mammals' genetic program. To provide a hypothesis about the common elements of individual preferences, something which is missing in economic theory, these items may as well be considered part of man's genetic endowment.

However, besides these innate, primary reinforcers, behavioral psychology suggests the existence of so-called "secondary" or conditioned reinforcers. They are acquired in a conditioning or learning process which builds on the innate learning capacity shared by all vertebrates including man (Skinner 1966; Pulliam/Dunford 1980, pp. 11-44). According to this theory, learning secondary reinforcers takes place by simple association with primary reinforcing events. If an originally neutral stimulus is systematically paired with a primary reinforcing stimulus, an association is established in such a way that the learned stimulus can eventually act as a substitute, at least for a limited time. A stimulus associated with conditioned reinforcers can also acquire reinforcing power in this way. It can easily be imagined, given man's large associative capacity, how an extremely developed hierarchy of learned reinforcers emerges from the very simple innate learning mechanism.

In the light of this approach, some instructive conclusions can be drawn with regard to human preferences. Their variety can obviously be explained as the result of a permanent learning process, a slow process of preference formation and change. In an individual lifetime history of conditioning, a more or less long chain of learned associations leads from the few innate preferences to those actually revealed in current economic choices. These choices do not necessarily increase genetic fitness -- people may, for instance, prefer to enjoy increased consumption rather than to raise children. Yet their decisions are related, in a complex, idiosyncratic fashion, to basic preferences that once had an adaptive value. In addition to the individual history of interactions of primary and secondary reinforcers, the unique historical circumstances that make certain choices available to certain individuals at a certain time also, of course, contribute to the enormous variance in actually observable choices in (and outside) the markets. The question that remains is whether, in spite of all the variance resulting from the individual influences, it can be argued that a general tendency can arise in the grand total of human choices in a way that is significant for characterizing economic evolution and, if so, whether such a tendency has something to do with the basic innate preferences of man.

V. Economic Evolution - Between Anthropocentrism and "Nature's View"

It has been argued in the previous sections that the Darwinian theory of natural selection is of little heuristic use for conceptualizing evolution in the economic domain. There may, however, be an indirect influence through some basic, innate, human preferences which play a role in organizing and directing the individuals' learning or preference formation processes. Before this question can be decided, it is useful to briefly outline some of the significant features of economic evolution which have been observed particularly over the past centuries. An almost trivial fact to be mentioned first is the undiminished exponential growth of the human population. An important prerequisite for the economic subsistence needed to feed a growing population has always been the provision of labor for the production of the means which, in turn, are capable of satisfying the basic needs. However, it is only in the last four centuries that growing technological knowledge made two things feasible at once -- a large scale substitution of human physical work through non-human energy sources and a real boost in the production volume. Wind and water power, wood, coal, oil, natural gas, uranium were tapped one after another.

Coincidentally, the price of energy per unit, measured, e.g., in Mega-Joules, fell strongly relative to the wage rate. This means that, on average, one hour of work was able to buy an increasing amount of energy. Furthermore, the mechanization of the transformation of materials at large increased output produced per hour worked, i.e. the general productivity of labor. Since, however, energy intensity grew faster than labor productivity, the energy input per unit of output, the 'energy coefficient', increased. This is particularly obvious in the agricultural sector where a direct comparison between 'primitive' methods of cultivation based exclusively on human energy inputs and modern methods based on non-human energy inputs is possible (see Pimentel/Pimentel 1996, Table 10.3, 10.4, and 10.7). Needless to say, the productivity increases in agriculture thus achieved were instrumental in feeding a growing number of people. In addition the productivity increases in the agricultural sector improved the nutrition of the existing population and resulted in an increasing life expectancy as recent research on the period since 1800 has shown (see Fogel 1986).

The strongly improved 'terms of trade' between human labor and nature's products (non-human energy flows and, with the necessary accumulation of machinery, materials) explain a fact that may seem puzzling, at least to non-economists. In their historical development, modern economies could afford to shift the bulk of their expenditures (or, to look at it the other way round, the bulk of their income creation) first from agriculture to the industrial sector and then on to the service industries. Nowadays, agriculture and the industrial sector together generate less than half of domestic income in all OECD countries. Because of the comparatively high input share of human work hours needed in producing services, these services have become relatively more expensive compared to agricultural and manufactured goods. At the same time, the demand for services does not seem to have been curbed by their increasing relative price in favor of an expansion of demand for agricultural and manufactured goods, perhaps because the demand for the latter goods is more quickly satiated than demand for services.

Indeed, another striking feature of economic evolution over the past few centuries turns up on the demand side of the markets, where mankind's consumption has grown continually both absolute and per capita. In both the developing and the developed countries this is accompanied by increasing differentiation of consumer goods and the generation of entirely new groups of products. Economic growth is no longer only a matter of capital accumulation, technical progress, and productivity increases. It is also a story of what can be sold in the consumers' markets. Qualitative change in the array of final consumption goods, increasing mobility, internationalization of trade in formerly local specialities, and the soaring diversification of the service sector are only the most salient aspects of a deep seated process of change. This change is not without regularities. Economists have long known that the demand for certain product groups, such as groceries, does not expand at the same rate as income available for consumption does -- a sign of satiation despite all the differentiation efforts in those industries. For other product groups, such as housing space and, in particular, those products which directly or indirectly provide comfort, safety, improvements in health, mobility and entertainment, the demand effects are exactly the opposite, i.e. they indicate that satiation is a long way from being reached. The latter products are, of course, precisely those provided by the economy's service sector.

Are all these empirical findings just accidental or do they fit into some general causal patterns? Economists are rarely inclined to accept this as a question they feel compelled to find an answer for. If long run economic development is considered at all from a theoretical point of view, it is usually treated as a process of economic growth that reflects an increase in the (subjective) utility which most of the agents involved enjoy (Romer 1986) -- the anthropocentric view of a basically hedonistic economic discipline. In an evolutionary perspective things may be interpreted differently. Such a perspective considers human kind as just another species, albeit one that has been able to enjoy a prolonged period of affluence, because its enhanced problem solving capacity just kept pace with the ecological problems created. As was explained in the previous section, human decision making can be seen as based on a complex structure of preferences that have emerged over the individuals' lifetimes from a few innate, and therefore basically inter-individually shared, dispositions. Those complex structures vary greatly and reflect the idiosyncratic learning history. It may therefore be very difficult to trace back the choices made by an economic agent at a given time to the few innate preferences that agent shares with other agents. Nonetheless, the only consistent and enduring effect that could, on average, be expected to show up despite all the individual variance in the decision making of the human kind over an extended period of time may be precisely an expression of those basic preferences that were fixed in the earlier times of human phylogeny.

Some of the features of the long term economic evolution sketched above do seem to fit this conjecture well. Humans behave much like any other species that dominates its environment. They increase their population size and pay little attention to the ecological dangers a rapidly growing population means, and they make efforts to enhance individual health and life expectancy. They subdue all rival species or drive them to extinction. Humans use their growing knowledge to find ways of avoiding the burden of physical work in making a living. As far as they can, they expand their consumption of primary reinforcing items -- food, drinks, entertainment, mobility, living space, and so on -- to a level of satiation. Before satiation is reached, suppliers of those items search for and create more sophisticated ways of consuming those items. The satiation features of some items, such as food, set more narrow limits on the expansion of the consumption than others like the diverse services mentioned above. Since many of these services, in particular those relating to mobility, rely more heavily on energy flow inputs than on material inputs, energy consumption is unlikely to reach satiation in the future. If the working hypothesis about the role of innate primary reinforcers or preferences suggested above holds, it may be argued that a Darwinian perspective on economic evolution is fruitful, and that it leads to important material conjectures about economic evolution -- provided, however, that perspective does not build on misleading metaphors or misconceived applications of the theory of natural selection to the domain of economics.

VI. Conclusions

Human economic activity has changed significantly over time and, apparently, is continuing to do so in an accelerating fashion. In view of this, the static (and latently normative) interpretation of the economy in modern economic theory is quite inapt. An "evolutionist revolution" does seem warranted. However, unlike the amazingly simple and yet powerful principles of genetic variation and natural selection which, as Darwin discovered, underlie the evolution of the species, the regularities which economic evolution may follow seem to operate in a way that is far more complex. The idea of applying the theory of natural selection directly to explaining the rapid adaptations in modern economic behavior cannot be upheld. There is simply no evidence for a systematic, differential, reproductive effect of certain kinds of economic activities. Metaphorical use of core notions of the Darwinian theory -- selection, variation, and replication -- in an economic context is no less problematic. Economists sometimes indulge themselves in speaking of selection forces metaphorically to highlight the outcome of competitive processes, particularly market processes. Agents in the markets compete with each other and the relatively inefficient, incompetent, or simply hypertrophic, competitors lose and eventually go out of business (an economic, not a physical, extinction). But the differences between the actual economic and biological processes referred to by the term selection, and the results produced by these processes, are so very different that the metaphor is of limited value for economic theory. The simple reason is that, in the domain of economics, there is no structure comparable in its continuity with the genetic mechanisms that have led to the emergence of species in nature. An inspection of long term economic development can, however, help identify certain regularities which can be explained by evolutionary concepts as long as these are defined in a broader sense. A theory of preference formation has been suggested that points to some crude, genetically coded, human dispositions which may have a weak but persistent effect on economic choices and would explain the observed regularities. Economics may thus be fruitfully approached on the basis of a Darwinian world view in which natural selection has been, but no longer is, the immediate shaping agent.

References

ALCHIAN, A.A.: "Uncertainty, Evolution, and Economic Theory", *Journal of Political Economy*, 58 (1950), pp. 211-221.

ALLEN, P.: "Evolution, Innovation, and Economics", in: G. DOSI, C. FREEMAN, R. NELSON, G. SILVERBERG, L. SOETE (Eds.): *Technical Change and Economic Theory*, London (Pinter Publishers) 1988, pp. 95-119.

ANDERSON, E.S.: *Evolutionary Economics: Post-Schumpeterian Contributions*, London (Pinter) 1994.

AUDRETSCH, D.B.: *Innovation and Industry Evolution*, Cambridge, Mass. (MIT Press) 1995.

BANDURA, A.: *Social Foundations of Thought and Action - A Social Cognitive Theory*, Englewood Cliffs (Prentice-Hall) 1986.

BOULDING, K.E.: *Evolutionary Economics*, Beverly Hills (Sage Publications) 1981.

BOWLER, P.J.: *Evolution - the History of an Idea*, Berkeley (University of California Press), rev.ed. 1989.

BOWLER, P.J.: "Herbert Spencers Idee der Evolution und ihre Rezeption", in: E.-M. ENGELS (Ed.): *Die Rezeption von Evolutionstheorien im 19. Jahrhundert*, Frankfurt (Suhrkamp) 1995.

BOYD, R., RICHERSON P.J.: *Culture and the Evolutionary Process*, Chicago (University of Chicago Press) 1985.

DESMOND, A. and MOORE J.R.: *Darwin*, London (Michael Joseph) 1991.

EDGEWORTH, F.Y.: *Mathematical Psychics*, London 1881.

FABER, M./PROOPS, J.L.R.: *Evolution, Time, Production and the Environment*, Berlin (Springer) 1990.

FOGEL, R.W.: "Nutrition and the Decline of Mortalities since 1700: Some Preliminary Findings", in: S.L. ENGERMAN, R.E. GALLMAN (Eds.): *Long-Term Factors in American Economic Growth*, Chicago (Chicago University Press) 1986, pp. 439-555.

FRIEDMAN, M.: The Methodology of Positive Economics, in: M. FRIEDMAN: *Essays in Positive Economics*, Chicago (University of Chicago Press) 1953.

GHISELIN, M.T.: "The Economy of the Body", *American Economic Review*, Vol. 68, Papers and Proceedings (1978), pp. 233-237.

GOWDY, J.: *Coevolutionary Economics: The Economy, Society and the Environment*, Boston (Kluwer) 1994.

GÜTH, W./YAARI, M.: "Explaining Reciprocal Behavior in Simple Strategic Games: An Evolutionary Approach", in: U. WITT (Ed.): *Explaining Process and Change - Approaches to Evolutionary Economics*, Ann Arbor (University of Michigan Press) 1992, pp. 23-34.

HANSSON, I./STUART, C.: "Malthusian Selection of Preferences", *American Economic Review*, 80 (1990), pp. 529-544.

HAYEK, F.A.: "Notes on the Evolution of Systems of Rules of Conduct", in: F.A. HAYEK: *Studies in Philosophy and Economics*, London (Routledge and Kegan Paul) 1967, pp. 66-81.

HAYEK, F.A.: "The Three Sources of Human Values", Epilogue to *Law, Legislation, and Liberty*, London (Routledge) 1979.

HAYEK, F.A.: *The Fatal Conceit*, London (Routledge) 1988.

HERMANN-PILLATH, C.: "A Darwinian Framework for the Economic Analysis of Institutional Change in History", *Journal of Social and Biological Structures*, 14 (1991), pp. 127-148.

HIRSHLEIFER, J.: "Evolutionary Models in Economics and Law", *Research in Law and Economics*, 4 (1982), pp. 1-60.

HODGSON, G.M.: *Economics and Evolution: Bringing Life Back into Economics*, Cambridge (Polity Press) 1993.

HODGSON, G.M.: "The Evolution of Evolutionary Economics", *Scottish Journal of Political Economy* (1995).

KHALIL, E.L.: "Economics and Biology: Eight Areas of Research", *Methodus*, 4 (1992), pp. 29-45.

LEA, S.E.G.: "The Analysis of Need", in: R.L. MELLGREN (Ed.): *Animal Cognition and Behavior*, Amsterdam (North-Holland) 1983, pp. 31-63.

MARSHALL, A.: *Principles of Economics*, 8th edition, London (MacMillan) 1938.

MATTHEWS, R.C.O.: "Darwinism and Economic Change", in: D.A. COLLARD, D.R. HELM, M.F.G. SCOTT, A.K. SEN (Eds.): *Economic Theory and Hicksian Themes, Oxford* (Clarendon Press) 1984, pp. 91-117.

MAYR, E.: *One Long Argument*, Cambridge (Harvard University Press) 1991.

METCALFE, S. (1988): "The Diffusion of Innovations: An Interpretative Survey", in: G. DOSI, C. FREEMAN, R. NELSON, G. SILVERBERG, L. SOETE (Eds.): *Technical Change and Economic Theory*, London (Pinter Publishers) 1988, pp. 560-589.

METCALFE, S. (1989): "Evolution and Economic Change", in: SILBERSTON (Ed.): *Technology and Economic Progress*, London (MacMillan) 1989, pp. 54-85.

MILL, J.S.: *Principles of Political Economy With Some of Their Applications to Social Philosophy*, London (Parker) 1848.

MILLENSON, J.R.: *Principles of Behavioral Analysis*, New York (Prentice Hall) 1967.

MIROWSKI, P.: *Against Mechanism - Protecting Economics From Science*, Totowa, N.J. (Rowman & Littlefield) 1988.

NAKICENOVIC, N., GRÜBLER, A. (Eds.): *Diffusion of Technologies and Social Behavior*, Berlin (Springer) 1991.

NELSON, R.R. and WINTER, S.G.: *An Evolutionary Theory of Economic Change*, Cambridge, Mass. (Harvard University Press) 1982.

PARETO, V.: *Cours d'èconomie politique*, Lausanne 1896.

PENROSE, E.T.: "Biological Analogies in the Theory of the Firm", *American Economic Review*, 42 (1952), pp. 804-819.

PIMENTEL, D., PIMENTEL, M. (Eds.): *Food, Energy, and Society*, rev. ed., Niwot, CO (University of Colorado Press) 1996.

PULLIAM, H.R. and DUNFORD, C.: *Programmed to Learn: An Essay on the Evolution of Culture*, New York (Columbia University Press) 1980.

RICHARDS, R.J.: *The Meaning of Evolution: The Morphological Construction and Ideological Reconstruction of Darwin's Theory*, Chicago (Chicago University Press) 1992.

ROGERS, E.M.: *Diffusion of Innovations*, 3rd ed., New York (Free Press) 1983.

ROMER, P.M.: „Increasing Returns and Long-Run Growth", *Journal of Political Economy*, 94 (1986), pp. 1002-1037.

SAVIOTTI, P.P.: "Information, Variety and Technoeconomic Development", *Research Policy*, 17 (1988), pp. 89-103.

SAVIOTTI, P.P., METCALFE, J.S.: "Present Developments and Trends in Evolutionary Economics", in: P.P. SAVIOTTI, J.S. METCALFE (Eds.): *Evolutionary Theories of Economic and Technological Change*, Chur (Harwood Academic Publishers) 1991, pp. 1-30.

SCHMITT, G.: *Die Comptoise Uhr*, Villingen (Müller) 1983.

SCHUMPETER, J.A.: *The Theory of Economic Development*, Cambridge, Mass. (Harvard University Press) 1934 (1912).

SKINNER, B.F.: "Operant Behavior", in: W.K. HONIG (Ed.): *Operant Behavior - Areas of Research and Application*, New York (Meredith) 1966, pp. 12-32.

SMITH, A.: *The Theory of Moral Sentiments*, London/Edinburgh (Millar, Kincaid, and Bell) 1759.

SMITH, A.: *An Inquiry into the Nature and Causes of the Wealth of Nations*, London (Strahan and Cadell) 1776.

TRIVERS, R.L.: "The Evolution of Reciprocal Altruism", *Quarterly Review of Biology*, 46 (1971), pp. 35-57.

USHER, A.P.: *A History of Mechanical Inventions*, Cambridge (Harvard University Press) 1954.

VEBLEN, T.B.: "Why Is Economics Not an Evolutionary Science?" *Quarterly Journal of Economics*, 12 (1898), pp. 373-397.

WALRAS, L.: *Elements of Pure Economics - On the Theory of Social Wealth*, Homewood, Ill. (Irwin) 1954 (1874).

WEDERVANG, F.: *Development of a Population of Industrial Firms*, Oslo (Universitetsforlaget) 1965.

WINTER, S.G.: "Economic 'Natural Selection' and the Theory of the Firm", *Yale Economic Essays*, 4 (1964), pp. 225-272.

WITT, U.: "Economic Behavior and Biological Evolution - Some Remarks on the Sociobiology Debate", *Journal of Institutional and Theoetical Economics*, 141 (1985), pp. 365-389.

WITT, U.: "Economics, Sociobiology, and Behavioral Psychology on Preferences", *Journal of Economic Psychology*, 12 (1991), No. 4.

WITT, U.: "Evolutionary Concepts in Economics", *Eastern Economic Journal*, 18 (1992), pp. 405-419.

WITT, U.: "Moral Norms and Rationality within Populations: An Evolutionary Theory" in: J.C.PARDO, F.SCHNEIDER (Eds.): *Current Issues in Public Choice*, Aldershot (Edward Elgar) 1996, pp. 237-256.

YOUNG, R.M.: *Darwin's Metaphor: Nature's Place in Victorian Culture*, Cambridge (Cambridge University Press) 1988.

Conclusion

Chapter 14

The Theory of Evolution as Sociobiology and Bioeconomics A Critique of Its Claim to Totality

PETER KOSLOWSKI

Sociobiology constitutes a new development among the science disciplines on the frontier between the social and the natural sciences.[1] According to

---

* This paper is the revised and enlarged English translation of P. KOSLOWSKI: "Evolutionstheorie als Soziobiologie und Bioökonomie. Eine Kritik ihres Totalitätsanspruchs", in: R. SPAEMANN, P. KOSLOWSKI, R. LÖW (Eds.): *Evolutionismus und Christentum*, Weinheim (Acta Humaniora, VCH) 1986, pp. 29-56. - In some parts, the paper is based on P. KOSLOWSKI: *Evolution und Gesellschaft. Eine Auseinandersetzung mit der Soziobiologie*, Tübingen (Mohr Siebeck) 1984, 2nd edition 1989. English translation: P. KOSLOWSKI: *Ethics of Capitalism and Critique of Sociobiology.* Two Essays with a Comment by James M. Buchanan, Berlin, New York, Tokyo (Springer) 1996. The discussion of sociobiology from the point of view of the social sciences and economics is further expanded there.

1 WILSON (1975) and (1980), LUMSDEN/WILSON (1981).

the program of its founder, E. O. Wilson, from whom the very concept of sociobiology stems, it should effect a new synthesis between biology and the social sciences.

I. The Concept and Program of Sociobiology

Just as the science of physics serves as a forerunner to chemistry and chemistry to biology, the evolution theoretical science of biology has become, according to Wilson, the science underlying the social sciences and furnishes them with interpretative models and fundamental theories of social behavior derived from genetics and behavioral research. Out of the simple observable phenomena of behavior in the animal world, which can be viewed as strategies arising from natural selection that maximise survival, the complex forms of human social behavior are explained as and attributed to being basic functions of the maximising of genetic survival.

Along with this very comprehensive program of a human sociobiology, however, there is also a more limited empirical program of research intended by the idea of a sociobiology. As a part of biology, this sociobiology deals above all with the phenomena, functions, mechanisms and the genesis of social behavior among animals and attempts to explain this behavior by genetic-evolutionary and ecological-evolutionary theories.[2] The behavior of animals in their ecological environment, their interaction with individuals of their own and of other species, with competitors for food, with those who prey on them or with natural enemies, - all this is explained as a result of adaptations to the environment and of genetic selection of such individuals as those, whose strategies and adaptations maximise survival. "Veterinary sociobiology" is descriptive of a theory of interpretation of animal behavior, in which various theoretical and empirical methods are at work; game-theoretical methods and methods from economic and optimisation theory.[3]

Common to these approaches is the fundamental starting point of neo-Darwinian evolution theory that "nothing in biology makes sense other than

2 Cf. FISHER (1930), HAMILTON (1964), TRIVERS (1971), ALEXANDER (1974), GHISELIN (1974a), MAYNARD SMITH (1976), DAWKINS (1978).

3 MAYNARD SMITH (1976) applies game-theory models, GHISELIN (1974a) micro-economic and DAWKINS (1978) optimisation-theoretic.

in the light of evolution".[4] Evolution is understood as a process, in which those individuals are selected, whose behavior is best adapted to their environment and whose genetic fitness and reproductive success are, as a consequence, the greatest. The genetic information of the "fittest" maintains itself most comprehensively in the course of evolution. Life-forms do not pursue the "goal" of sustaining the species or the group, but solely the "reproductive imperative" of sustaining their own genetic information.

They seek to maximise their own fitness and the fitness of their immediate descendants (inclusive fitness), whereby priority in ensuring the survival of related stock correlates with the closeness of the degree of kinship.[5] Sociobiological approaches rule out group selection of any kind. The members of a species act solely for the benefit of themselves or of their direct kin (individual or kin selection). The behavior of the life-form is interpreted as being the result of strategies which maximise self-interest. There is an explanation and even a prognosis of behavior possible in precisely specified environmental contexts in reconstructing strategies, which make possible the most comprehensive and most enduring storage of genetic information. Sociobiology shows under which conditions specific strategies of interaction and work-sharing among animals are carried through in different environmental situations. It reconstructs under which conditions of the constituting of a species and its environment it is to be expected that differing forms of reproduction will arise such as hermaphroditism or bisexuality[6] or varying degrees of the division of work between castes, for example among social insects (worker-bees, drones, queens).[7]

A behavior theory, which begins with the assumption that all creatures follow the imperative of reproduction and of maximising inclusive genetic fitness, naturally carries this over to the field of human behavior. In accepting a uniform evolutionary process and the validity of genetic-physiological laws for human beings as well, veterinary sociobiology has to extend itself to human sociobiology. A transfer of this nature is represented in two radically different approaches.

---

4 Thus TH. DOBZHANSKY, quotes from MARKL (1980b), p. 1.

5 Cf. HAMILTON (1964), BECKER (1976), DAWKINS (1978), HIRSHLEIFER (1978).

6 Cf. GHISELIN (1974a).

7 Cf. OSTER/WILSON (1978).

The first and most venturesome approach to a human sociobiology (Wilson, Lumsden/Wilson and Trivers)[8] accepts that the same laws apply in human society as in the animal world. According to this account, all human societies follow the imperative of reproduction and of maximising the inclusive genetic fitness. Social institutions can be reduced wholly to being efficacious means to this end and they can be interpreted exhaustively in this way (reductionism). Intellectual and social phenomena are mere epiphenomena of the genetic-physiological basis of gene maximisation. The realm of the social and the cultural is thoroughly directed to the biological purposes. The cultural and social norms are to be understood as being the product of a biological evolution.

The second type of sociobiology, as represented by the most empirically oriented researchers, is more reserved in its theoretical claims. Its supporters concede a distinction between animal and human sociobiology and assume a plurality of fields of phenomena, from a difference between genetic-physiologically conditioned phenomena and phenomena determined by the conceptual and linguistic capabilities of the human spirit.[9] Ghiselin maintains that, in the sociobiology both of humans and of the animal world, the reproductive imperative and the laws of genetic competition and of competition for resources for reproduction or procreation are in force. According to him, however, the human spirit brings a new quality into human society, since it takes the edge off the competition in reproduction. It diverts individual energies over to such goods as those where no rivalry in consumption dominates. With these goods, reduction in the use of resources by the one person is not a reduction in the stock of those resources, which remain for another. The conflict-resolving and competition-reducing quality of art, religion and play mitigates, in human society, the violence of the rivalry for scarce resources in reproduction and establishes that animal and human sociobiology cannot be identical.[10]

Sociobiology is not a uniform methodological approach but a program with various ontological and epistemological starting positions. This program can be represented as a reductionist-monistic as well as materialist-evolutionist theory of totality, with claims to being a weltanschauung. It can,

---

8 According to TRIVERS (quoted in *Time* 110 [1976], p. 54), sociobiology will one day overtake the disciplines of "political science, law, economics, psychology, psychiatry and anthropology".

9 MARKL (1980a), p. 8. Cf. also MARKL (1976), pp. 25 and 36 and BUCKLEY (1977).

10 GHISELIN (1974a), p. 262f.

however, also be based on the pragmatic-hypothetical research program of the genetic-evolutionist biological analysis of animal and human societies, in which it is not yet known how far it will be productive in the field of human beings. The claim to justify an appropriate ontology of the social with evolutionist-genetic and evolutionist-ecological theories is then waived, or this ontology withheld until the empirical research has made further progress. The demarcation between sociobiology as a monistic "world-view" and sociobiology as a research-hypothesis and strategy of a comprehensive explanation of behavior is not always simple to draw on two grounds. For one thing, only a few authors such as Wilson and Lumsden explicitly declare their ontological basic tenets and their claims to complete explanation and they expressly champion the monistic ontology of a single process of evolution. Most authors renounce an ontologisation of sociobiology and the claim to a total explanation. They limit themselves to partial models of the sociobiological explanation of specified forms of human behavior, such as sexuality or territoriality.

For another thing, the thesis of the identity of animal and human social behavior and of the monism of the sociobiological explanation is introduced only indirectly by the use of metaphors from the human social field to describe and explain, in a sociobiological sense, the behavior of animals. To dispense with the disclosure and substantiation of the analogies between veterinary and human sociobiology is unsatisfactory, ontologically and epistemologically speaking. Since, however, social theories always have legitimising functions in a society and are a way of a society interpreting and projecting itself, it also has at the same time sociological-ideological side-effects. Social theories are not just descriptions but are, at one and the same time, interpretations and definitions of the social world, which are, or can be, self-realising in the sense of *a self-fulfilling prophecy.*[11] If the human society interprets itself according to the model of sociobiology and makes the reproductive imperative the ultimate purpose of its social teleology, this does not remain without consequences for the consciousness of its members and the character of the society. The introduction of social theory into social systems differs from the introduction of physical theories regarding extra-societal reality into a society. The repercussions of physical theories on the self-definition of a society are more slight than those of sociological and sociobiological theories.

11 Cf. THOMAS (1928), p. 81 and MERTON (1957).

II. Sociobiology as Bioeconomics

The ecology of evolution and sociobiology view living nature as a constantly changing relationship of adjustment and interaction, in which the life-forms, competing for scarce resources to maintain life and to procreate, have to make the most efficient use possible of these resources. The ecology is conceived of as an economy of Nature. In this bio-economy, the same laws apply as in the economy of humankind: "One universe, one economy, one economics".[12] Hence, theorems and laws of economic theory, especially those of microeconomics, can and must be drawn on to explain the evolution and the ecology of animal and plant behavior. For Ghiselin, the economy of Nature is throughout a competitive economy of a market economy type. Natural and sexual selection are a manifestation of the competition for scarce resources, as for example, for females capable of breeding. In this competition, only those, who make efficient use of their own resources, of their capital and of the resources of the environment, can survive and propagate themselves. Species and individuals, beasts of prey and their prey compete for food and for energy for their reproduction. The ecology is an economy, in which market laws apply. The extent of the market determines the scale of the division of labor and specialisation, the density of population on a territory the scale of the competitiveness and of the social differentiation in castes. The diminishing marginal returns of a production factor (law of diminishing marginal returns) fixes the optimum size of a firm, for example the size of a nest and the number of progeny, which are reared in it. The law of diminishing marginal returns from an activity and the raising of productivity through specialisation cause differentiation of genders and various forms of reproduction to arise.

Nature's economy is not an economy of linear optimisation or of simple maximisation of gene replications; nor is it an economy of frugality, but an economy of innovation and prodigal mutation. For sexual reproduction, i.e. the re-combination of the inherited DNA information from the genes of two individuals is, above all, wasteful. It uses up energy, material and time in the hunt for a sexual partner and disturbs the food intake and growth of the individuals. It is, though, a means of mobilising genetic variability and capability to adapt to changing environments. The DNA combination, constantly be-

12 GHISELIN (1974a), p. 212. Cf. RAPPORT/TURNER (1977) for a survey of the application of economic laws in the ecology.

ing altered by means of sexual reporduction, makes accommodation with the environment and the exploitation of new resources possible. According to Ghiselin[13], the flow of genes in the re-combined DNA is comparable to the flow of capital and technical knowledge, a flow, which leads capital and know-how into ever new productive uses and makes possible innovative reallocation of the existing capital stock.

According to sociobiology the aim of nature is maximum survival of the genes. Dawkins's statement presents the most radical impress of the monism of the optimisation of the survival of genetic information. Life-forms, according to Dawkins, are survival machines of the selfish gene. "We are survival machines - robots, blindly programed to preserve the selfish molecules, known as genes."[14] The genes furnish the life-form with an optimising program for mastering complex situations in the external world and provide for the safeguarding of an optimal internal and external environment for the ultimate purpose of the survival of the genes. This program regulates the self-replication of these genes over several generations.

Such a model subsumes the history of the human and of nature under one single purpose, the purpose of the survival of the genes. The task of the human through life and her purpose are solely the maximising of the survival of her genes; the end purpose of humanity is the maintenance of the gene pool. The history of humankind is no longer a continuance of natural history, but is identical with the history and purpose of nature, since it is solely a matter of preserving genetic information in both. Before the categorical imperative of gene survival, all other imperatives become hypothetical imperatives. To be consistent, this would lead to the justification of the cannibalism, such as Jünger feared as an outcome of every monistic-economic approach: "First steps, transitions to intelligent cannibalism are alluded to, often even undisguisedly. Every purely economic view has necessarily to advance to that."[15] The optimal use of food resources, according to sociobiology, requires that the animal mothers eat up those of their young, that do not serve the survival of the genes, for the benefit of siblings capable of survival.[16] To be sure, authors like Dawkins and Wilson[17] are averse to transfers of this kind from

13 GHISELIN (1974a), p. 54.

14 DAWKINS (1978), p. VIII.

15 JÜNGER (1958), p. 160.

16 DAWKINS (1978), p. 158ff., GHISELIN (1974a), p. 231.

17 DAWKINS (1978) is particularly inconsistent here. On the one hand, we are blind robots, the selfish genes make use of (p. VIII), while, on the other, we

the bio-economy to the human one. If cannibalism in families were to promote the survival of genes, this consequence would, nevertheless, have to be drawn from sociobiology. Wilson adheres to the rights of man, but his argument for them goes way beyond his own sociobiological concept.[18]

Dawkins's approach concedes a faculty for aspiration, intentionality and consciousness to the genes. In doing so, he falls into a genetic animism, which apportions perception and decision to the genes and oversubscribes by far to the efficiency and the speed of Darwinian selection mechanisms. Such a kind of animism leads to the ontologising of economics and of the concept of the rational pursuit of goals. The genes are seen as "investors of capital in a stockmarket" whose shares or enterprises, as the case may be, are the vehicles of survival, the bodies, in which these genes invest. The genes behave like profit maximisers in the programming of the construction of their survival vehicles; like entrepreneurs, they optimise the profit in their strategies and measure their "investment policy" against the "gold standard of evolution, the survival of the genes".[19] For Dawkins accordingly, the genes also possess a capacity to learn, which corresponds to that of individuals, who maximise profits. The difference of the time horizons of genetic and intellectual "learning" is mixed up and neutralised in animistic manner by Dawkins.[20]

---

are able to thwart the plans of the genes (p. 3) and cannot deduce, therefore, ethics from the laws and results of evolution.

18 WILSON (1980), p. 187.

19 DAWKINS (1978), p. 67 and p. 146f.

20 DAWKINS (1978), on pp. 56, 145 and more frequently, excuses his use of anthropomorphic criteria as an abbreviated way of speaking, which allows the complicated inter-relationships to be shown more simply: "Thus, it is often wearisome and unnecessary to be constantly bringing genes into it, when we discuss the behavior of survival machines. In practice, it is normally appropriate to regard the individual body approximately as a subject, which 'seeks' to augment the count of all its genes in future generations". Against this, it must be objected that the metaphors matter for the argument. Much more do they conceal the fact that the named anthropomorphic capabilities and subject qualities of genes cannot be reconstructed by theorems of a purely selection-theoretical nature. Much more do they allow work to be done, using teleological concepts while maintaining, nevertheless, the appearance of a thoroughly mechanistic model. Cf. LÖW (1980), p. 283, for a critique of the use of metaphors and quotation marks in modern biology.

III. On the Ontology of the Theory of Evolution

Dawkins's approach raises the ontological question of what we have to see as the ultimate basic principles of that which exists; the form and its realisation or the sustaining of genetic information. By shifting the phenomenon of intentionality into the gene, Dawkins maintains, that the conservation of the genetic data is the ultimate final cause of all that exists. The conservation of the genes stands well above the purpose of the preservation of the individual or even the species. It is not that the form of the organism is preserved as a program in its entirety but that the elements of the genetic codes aim at the augmentation of their capacity to survive, independently of the organism as a whole. "The rather egoistic large particle of chromosome and the even more egoistic small chromosome particle" strive for self-preservation.[21] Not only the concept of the species - as in Darwinism and Neo-Darwinism[22] - is dissolved in a nominalistic manner but the concept of the individual too is subjected to a genetic supernominalism. The identity of that which safeguards its genetic fitness is no longer determinable linguistically or empirically. It makes no sense to talk either of the history of the genus, humankind, or of that of one of the individual persons. One can only speak now of the history of the gene pools, which use such machines, that exhibit an astonishingly high degree of fidelity to form and are named by these machines themselves, using an artificial word, namely human. The concept of species and of individual have no point of reference; they are no longer indicative of something real.

"The human" is now just an all-embracing term for a symbiosis of machines, that serve to preserve the data held in small chromosome particles.

1. The Maintenance of a Program as Entelechy?

Dawkins carries the supernominalism of the breaking down of the individual to an aggregation of survival machines to the point of absurdity: Perhaps we are not individuals at all, but consist of multiple organisms, whose genes make use of our consciousness of identity, in order to safeguard, in a symbio-

21 DAWKINS (1978), p. 39.

22 Cf. MAYR (1967), p. 337 to the critique of it in FELLMANN (1977), p. 286.

sis, their preservation within us?[23] The genetic supernominalism, according to which it is neither species nor individuals, but small units of genetic information, which are the ultimate determinants of the actual and of the living, veers dialectically round to an abstract essentialism or super-idealism of a pseudo-platonic type.[24] The existence and the survival of small units of information, which search out their living carriers and "exploit" them, constitute for Dawkins the essence of everything living. On the other hand, the individual's nature, having a body and shape, becomes an epiphenomenon of the real being, the being of the genes.

Against such a genetic "idealism" is the argument, that suggests itself as to why something possessing shape should become that way and take on existence at all, if its teleological purpose is merely the survival of something quite different, invisible and without shape. If the survival of the genes is the purpose and this survival program directs the actuality of what is alive, then the actuality, perceptible by us, is in great measure non-functional or heterositic, since it is indeed possessed of shape and we humans too are the result of being shaped in materialised form and not the substrate of abstract idealisms. It would be more economical for the genes, to swim eternally in a primeval soup and to keep their data content in a state of potentiality without ever transforming this data into actuality with form. The *materialisation* of the DNA information in the shape of the individual is ontologically superfluous, if the purpose is merely the preservation of this information. It is superfluous and purposeless actually to build the cathedral, if it is just the architectural plan that is supposed to be preserved. To see in the genes the real perpetrator and to detect the *causa finalis* and the *causa efficiens* of all that exists in the individual gene is to confound causality with book-keeping.[25]

The reductionist definition of life by Dawkins as a replication of genetic information leads to the conclusion that his theory is implausible if viewed as optimising theory. Life and information could be sustained with less ex-

---

23 DAWKINS (1978), p. 213.

24 Cf. MASTERS (1979), p. 251: "For contemporary biology as for Plato, the visible animals can be said to 'represent' the information or *eidos*, which precedes them in time and space". For Dawkins, this turns out not to be true, because the *eidos* is divided into little pieces of chromosomes. The comparison by Masters of the genotype with the Aristotelian concept of the soul as *forma corporis* also turns out not to be true, since, indeed, it is precisely the stability of this form that has been relinquished by sociobiological theory.

25 Cf. GOULD (1984).

penditure, since the replication of information is carried out more expediently and more economically without actual organisms being involved. Why is it held that a human - or the genes working through her - generates a human[26] and not held that it is the small particles of chromosomes that do precisely the same? The concept of procreation has been thought of as implying that the being procreated has shape and as being related to species since the time of Aristotle, and it is for that reason not applicable to genes. Dawkins would have to be interpreted in the following way: a human begets a human, whereby a gene is replicated; the gene replicates itself in that it occasions a human to beget another human, who boasts of half the same genes. This is not an economical procedure. It corresponds to a printing shop, which, for the production of copies, always copies the copying machine at the same time and, in doing so, obtains copies, which only halfway correspond with the original. Replication and copying are not progenitive acts. Children are not copies or replications of their parents but their progeny, i.e. they are not identical with but are also not completely different from their parents. The relationship of descent is not a relationship of identity. Neither is it a relationship of the particular and of the universal or of element and class, since it is dependent on place and time. It is a unique relationship of the identity and non-identity. The universality of a species is passed on with the particularity of that organism by means of a begetting process. The organism is at the same time both singular and specific of a type. Hence, ordinary language also differentiates begetting from making, creating or copying. Identical copies, identical right down to the material foundations can be made and created but not begotten. Being mere imitations or duplicates, they are, also, as a consequence, of less value and "ontologically sterile".

The gene as an idea and as pure information substance, *in potentia,* can exist for eternity without corporeal substance, without laying itself open *in actu* to destruction in acquiring form. If, on the other hand, one comprehends the gene as entelechial substance, which has to become actual, then it is not the survival of the potential information that is the purpose of that, which is living, but the form and its actualisation. The inversion of the entelechy of the form to the survival of the genetic information of this form is not ontologically plausible and is self-contradictory. It supposes the gene, something being eternal and potential, to realise in something materialised and finite, the form of the soma. The end-purpose, however, is not the finite as something

26 Cf. ARISTOTLE, *Met.* Z., 1032a25: „The human begets a human."

finite and materialised, but the survival of the potential as potential. With the economic and selfish gene, the radical reversal of the teleology of the form to the teleology of the information is introduced to theory, the reversal from the accomplishing of the potential in the actual *(energeia)* to the maintenance of the potential as a potential *(dynamis)*. The potential, the realisable, strives to maintain itself as something potential and realisable, and not as something actual.

The question arises as to what genetic information can signify, if its materialisation with shape or its conscious representation for us, its entelechy in the form and shape accomplished or its being-for-another, for our consciousness, are not considered to be the final cause for its realisation. Information and ideative content are always only for a state of awareness and do not, in themselves, constitute mental content.[27] The genes are, of themselves, indifferent as to their realisation or non-realisation. Only for us does their realisation or non-realisation make a difference. Information is only conceivable and assertable as information made real or present. The genes cannot aim or "intend" their being made real, since their reality is potentiality. They are not the entelechy of that, which exists, but their taking shape in the realised form is the entelechial structure of reality.

In the wake of the three affronts to the naive self-love of humankind, so named by Freud, namely the affront caused by the change brought about by Copernicus, the affront of Darwin's theory about humankind's ancestry and the affront given by the psychoanalytical depotentialising of the Ego, Dawkins's affront would be the fourth and final one. It would not only drive humankind away from the centre of the Cosmos, rob humankind of her uniqueness among the living species and of her self-consciousness but also expose as a sham her physical individual character and her self-preservation. Dawkins's theory is the ultimate form of ontological nihilism. It is nihilism, because the being of the person is not acknowledged as being worth preserving for itself. Life serves merely to preserve something, that is itself not actual but only potential. The reason for being is the preservation of a program.

Dawkins's theory has not remained uncontradicted within biology. The criticism has been made, that it is not the gene, but the organism as a whole that is the basic unit on which selection is effective. The organism, or the *soma,* is the object of the selection, because the gene for a particular organ, for example for the liver, has little chance of survival, if it wants to assert

27 Cf. SPAEMANN/LÖW (1981).

itself individually against the organism.[28] The selfish gene, which is in competition with other genes of the same organism in the same organism, remains a construct. Dawkins's ontological nihilism is too much in conflict with the biological significance of the body, the *soma*, to be plausible, biologically speaking.

2. A Critique of the Optimisation Theoretical Reconstructions of Evolution

The model of the selfish gene unifies biology and economics, in that it relates all actions of living beings, functionally, to the maximising of inclusive genetic fitness. The objection must be raised against these optimising models in evolutionary ecology, that reconstructions or as-if-explanations of optimisation have the character of ad-hoc-explanations. They are ad-hoc-explanations to the extent that the cohesion of the explanation is defined more or less ad hoc and arbitrarily. The complexity of the situation, which an economical gene confronts, must of necessity be so greatly simplified in the theoretical optimisation reconstruction, that the description of the situation as antecedent conditions of the explanation bears only a distant resemblance to reality. "A simple theory concerning phenomena, which by their nature are complex, is probably of necessity false, at any rate, without a specific *ceteris paribus* clause, when, if completely formulated, the theory would no longer be simple".[29] The same points of criticism hold good for the reconstruction of the genetic selection as a process, that optimises survival of the gene, as for the economic models of maximisation. If they are to portray reality, these models require knowledge, that for the most part is not available.

In order that an optimising or, as the case may be, a maximising model may accurately describe reality, the following conditions must be fulfilled:

Firstly: the initial situation must be accurately described.

Secondly: the number and type, the "set", of the possible strategies must be known.

Thirdly: the criterion for the optimising, i.e. the variable, which has to be maximised, must be well defined.

28 Cf. GHISELIN (1981), pp. 305, 309 and (1984).

29 HAYEK (1972), p. 16.

Fourthly: the constraints of the maximisation process are to be precisely determined.[30]

These conditions are fulfilled only for "well structured" decision problems, which are capable of being clearly demarcated, such as those of technical optimisation in engineering science. Already for complex economic situations, such as the management of an enterprise, the maximisation model can no longer be a straightforward solution or guide to action because of the uncertainty over the strategies of others and on account of the complexity of the environmental conditions and of incidental side-effects.[31]

For the reconstruction of biological populations according to the rules of gene optimisation, the objections made by decision-theory are yet more significant in regard to the definition of initial situations and the effects of actions. In selection models, no rational forecast is given, the gene's decision situation is externally delineated by the observer and the constraints and strategies of the gene are not given or fixed, but they are reconstructed by the same observer.

Additional to these decision theory and economic objections appear others that are biological in a narrower sense. In biology, the arguments concerning optimisation start from the premise, that, if a specified phenotype has adjusted optimally, the appropriate genotype corresponding to it comes into being, which is something that has not been established. For some optimal solutions are mixtures of phenotypes, which are dynamically possible only under extremely restrictive genetic conditions. Lewontin summarises the problem convincingly: The dynamics of natural selection do not include rational foresight and there is no theoretical principle, that secures optimisation as a consequence of selection.[32] In biology, arguments for optimisation are applicable only with great caution, if they are not to be an inadmissible carrying over of anthropomorphic arguments into the theory of selection. Arguments for selection and for optimisation cannot both be valid at the same time. Either the selection model is valid, in which case it cannot be derived from optimisation with quasi-complete foresight, or the optimisation model prevails, in which case genes have to be thought of as having intention and foresight and Nature has to be considered as leading, teleologically, towards opti-

---

30 Cf. to the critique of optimisation models OSTER/WILSON (1978), p. 297 and LEWONTIN (1979), p. 8f.

31 Cf. ALCHIAN (1950), SIMON (1978) and KOSLOWSKI (1982), p. 230ff.

32 LEWONTIN (1979), p. 12.

mal development. Selection would only ensure, in a strictly optimisation theoretical model, that an equilibrium between the optimisation strategies of the kinds comes about in the same way, as the market operates between the self-interested strategies of rational economic individuals, according to the theory of the general market equilibrium of economic theory.

It is not possible to declare optimal solutions for the universal context of nature. The theory of the selfish gene in the working basis of sociobiology is not applicable as a theory of the evolution of the totality of living species across the whole spectrum. The optimisation criterion is already unclear. Is the continuance of parts of chromosomes optimised (as with Dawkins) or the continuance of whole chromosomes?[33] Is it the current inventory of genes or an intertemporal gene stock that is maximised?

On account of gaps in our knowledge and of the exponential growth in complexity of larger systems, we can state only the optima for regional contexts, so-called local contexts. The genesis of a whole species cannot be understood as optimisation. It may only be formulated thus: if populations or individuals, in a given environment under given incidental conditions, seek to safeguard the survival of their genes, they should behave in such and such a manner. How it gets to the point that an individual, as phenotype or genotype, finds itself in such a situation, requiring decision, cannot be represented afresh as a result of preceding optimisations; neither can it be reconstructed in theory. The optimising theoretical description or reconstruction of the evolution of a species is either incomplete, because we lack the necessary information or it leads to an unending regression of reconstructions of the present via reconstructions of the past.[34]

If one, however, comprehends the whole process of the evolution of a genus as optimising adaptation, this presupposes the *a priori* agreement, that adaptation leads towards optima, before any experience and subjection to em-

33 Cf. ALEXANDER (1974), p. 374: "What is it, after all, that selection is maximising?".

34 Here, in an inadmissible totalising and universalising, the empirical research model is treated speculatively as a holistic principle. According to KRINGS (1984), this transition from a partial empirical theory to a universal principle, this *pars pro toto*, is a mistake in logic of the type "metabasis eis allo genos" and marks the point, where empirical research into evolution is transformed into the metaphysics of evolution and natural science theory unintentionally encroaches onto philosophy's terrain and becomes a metaphysics.

pirical testing. It presupposes a teleology of nature, whose teleologism exceeds the teleology of the Aristotelian ontology of the natural kinds.

What is true of Darwin's theory of selection holds good also for the optimisation argument: both relate, above all, to the relation of the organism to its environment. Both emphasise the external aspect of adaptation of the organism to given conditions - *ceteris paribus* -, but are unable to reconstruct *in its totality* the internal aspect of the genesis of a genus and the genesis of the "setting", in which a genus or individual finds itself; both can only make assumptions regarding this.[35] The selection principle, if it is supposed to serve as a principle explaining the genesis of the reality of living matter as a whole, starts from the *a priori* assumption, that the genesis of the species is also to be explained by selection there, where the selection processes can no longer be reconstructed. From the observable and reconstructable selection processes, the theory of selection reveals that all existing things arise as a product of similar selection processes. This inductive inference is only possible when it is taken, *a priori*, that all selection operates in the sense of the pres ervation of adaptations. The Neo-Kantian school has, therefore, interpreted the theory of selection as a case of empirical application of the transcendental principle of teleology, that Kant has defined as "purposefulness of a composition of the diverse that is, taken in itself, accidental". "Natural selection explains in a mechanical manner the existence of forms that have proved to be fit. The principle of selection, namely, judges all processes of evolution from the standpoint of utility in the sense of their usefulness for conservation. It turns out, therefore, to be entirely accurate, when Darwin is characterised as 'the greatest teleologue of all'".[36]

Since reconstructions of the genesis of an organism, be it by selection theory be it by optimisation theory, are possible only as partial and not as total reconstructions of the history, the trans-specific evolution, overlapping the species, is not made clear in selection theoretical terms.[37] Selection theory, for sure, denominates, speaking in game theory terms, the "operative"

35 Cf. PETERS (1972), p. 349.

36 FELLMANN (1977), p. 294.

37 According to LEWONTIN (1978), p. 159, adaptation does not improve the chances of an organism's survival but compensates only for environmental changes and acts so that the organism keeps pace with these changes. Trans-specific evolution, however, is not to be interpreted as keeping pace with environmental changes through adaptation only since something new is beginning.

rules of play and the successful game-strategies of evolution. It cannot "account for" and reconstruct the starting genetic endowment and the total present genetic endowment as well as the "*status quo point*"[38] of the species engaging in the selection competition against one another.

3. The Impossibility of a Total Reconstruction of Evolution

Wilson and Lumsden/Wilson[39] see the efficiency of a scientific theory in its reducing the greatest number of phenomena to simple concepts in an aesthetically satisfying manner. For them, experience of natural science has shown this to happen best in the way that the real world is seen in a matrix of possible worlds.[40] Dawkins makes this program of possible worlds his own, in that he reconstructs his evolutionary stable strategies on the model of the best of all possible behavior patterns. Sociobiology puts in the claim to create a monistic theory of universal reality as outcome of a uniform evolutionary process and in reconstruction according to the model of possible worlds. Two objections may be raised against this:

It makes no sense to speak of possible worlds, if asking about the explanation of a process that is *per definitionem* unique, namely evolution as history of the universe. It is not admissible, in answer to the question of how this unique process can be reconstructed, to refer to possible worlds and simulations of paths of development or partial developments, as the case may be, since it is not possible worlds, which are of interest to us, but a single world, namely the real world and its development. Eigen/Winkler[41] admit that "only selected primal happenings can be experimentally examined, not however the historic chain of events of evolution". The development as a whole of the real world in all explanatory theories, involving selection, optimisation and game theory remains inexplicable in the end. The exclusion of antecedent conditions in explanatory models, through which what is irrelevant is eliminated or constraints are provisionally presupposed, is not legitimate for the examination of the sum and of the totality of the process. If the world system as a whole is supposed to be interpreted, then no "environment" can be excluded

38 Cf. LUCE/RAIFFA (1957), p. 124f.

39 WILSON (1980), p. 18, LUMSDEN/WILSON (1981), p. 346.

40 LUMSDEN/WILSON (1981), p. 2.

41 EIGEN/WINKLER (1975), p. 195.

as irrelevant. The attempt to retrieve the antecedent conditions of an explanation and to define the system in its entirety, leads, however, into an endless regression.

Secondly, it makes sense only for part-worlds, related to other part-worlds, to have, as a starting-point, possible worlds - or even the best possible world - as a heuristic model. World and evolution are possible or contingent only for a divine consciousness, and evolution could have taken a different course only for such a consciousness. For a finite consciousness, evolution and the world as a whole are as they are. Only for part-worlds, a finite consciousness is capable of judging whether they are optimal in respect to well-defined purposes. An absolute and infinite consciousness alone has the power to say that the world is the best among others, conceivable or possible. To apply the optimisation argument to the theory of evolution as a total outcome of optimising genes in the aggregate, leads either into a super-teleologism of the genes as eternal ultimate causes or into a theodicy, which interprets the evolution as the development of the best of all possible worlds. Both positions are incompatible with a causal-mechanical theory of evolution. The problem of the reconstruction of sociobiological behavior with the aid of possible optimal strategies proves as long the incompatibility of the selection theoretical and optimisation theoretical arguments as the theorem that selection leads to an optimal state or that the optimal state is selected is not proven. As-if-explanations and reconstructions, following arguments of optimisation, such as Dawkins employs, are not causal explanations in the strict sense, but plausible stories about what genes should do, if they want to ensure their survival. As plausible stories, they have heuristic value.

IV. Natural Kinds as Individuals and the Infinite Definedness of the Individual

The objection of the impossibility of reconstructing the totality may be raised not just against the optimisation theory. It holds good, too, for other approaches to the evolution theory's explanation of the formation of the species and of the origin of the species. Ghiselin has put forward the thesis, that species are not classes of organisms or aggregates of physical individuals with like characteristics or features. They cannot be classes, since they are not dependent on time or place. They are much more individuals in the logical sense of "particulars", combined complex unities or "composite wholes".[42] Species are united by a common line of descent. No genetic mutation, however great, can transform one species into another, since one type will always have the same line of descent and, logically, the same reference and the same designator. Ghiselin substantiated his proposition, that the species be viewed as individuals and not as classes, by means of two arguments. The perception of the species as a class or as of a natural kind excludes it being possible to bear in mind evolutionary mutations of this class. The class must remain identical in its defining attributes through time. According to Ghiselin, though, evolutionary change of the species does occur. Therefore, the species must be viewed as a mutating individual, which is defined not by its attributes but by the coherence of its lineage.

The second argument is served by an analogy from economic theory: Not only does competition for resources building-up the *soma* and for reproduction between the species occur, but it occurs also between members and organisms of a species. Trans-specific competition and infra-specific competition must be comprehended and explained by the same principles. The species, consequently, should be imagined, in Ghiselin's theory, as being analogous to business firms: the members of a firm compete with one another within the firm and the firms in their turn compete with other firms. The genera then form the sectors or branches of the economy of life. The theory of nature being a competitive economy implies that life constitutes a unity, whose parts are in a state of competition in various sectors (genera), firms (species) and organisms. Life becomes a "monophyletic whole"[43], that develops into the individuals of the species and, further, into the constituent parts of these individuals, the organisms.

If the species are to be seen as individuals, their designations must be proper names for a particular thing. Proper names and expressions for natural kinds are, however, "rigid designators", as more recent analytic philosophy has indicated (by S. Kripke among others). A designator is rigid, when it refers to the same thing - or to the same things - in each and every possible world, in which it has a reference. A rigid designator has the same reference in

42 Cf. GHISELIN (1974b) and (1981).

43 GHISELIN (1981), p. 281: "The ontological status of life is itself a very difficult problem. The usual approach is to treat it as the class of organisms, but an evolutionary systematist might consider it as a monophyletic whole, and hence an individual."

every contrafactual situation, in which it has any reference at all.[44] Names are attached, as it were, to individuals through a causal indissoluble chain, which does not denote what is common to a concept or a class, but presents the attachment of a name to an object and has a bearing solely on this object. It is almost impossible to determine a species by means of general characteristics. Only combinations of characteristics can be specified and, in the event, only the identification by pointing to the matter is all that remains, the pointing to the object as the renewal of giving the name as a rigid designator to the object.[45]

Defining the species as individuals with proper names, their presentation as rigid designators, corresponds to the description of the account in the Book of Genesis, that it is the task of humankind to give names to living creatures. According to *Genesis*, Chap 2, Verse 19, "God brought the beasts of the earth to Adam, to see what he would call them; for whatsoever Adam called any living creature, the same is its name." The naming of the creatures by humankind is, according to the Bible, the human's central contribution to the Creation.

Ghiselin's theory concerning species as individuals and evolution as interaction among these individuals exhibits great merit in contrast to the definition of species as classes. Ghiselin sees the provenance of the overestimation of the concept of the classes as stemming from the influence of Russell's logical positivism, in which what is real dissolved into relations and classes. Ghiselin's thesis, on the other hand, allows the unity of what lives and the homogeneity of infra- and trans-specific competition to become comprehensible. One thing, though, it does not accomplish - contrary to Ghiselin's opinion: it is not an explanation of the origin of the species. The theory of the species as individuals enables the mutation of the species better to be explained than does the theory of the species as classes, but it cannot reconstruct the origin of the species. The complete reconstruction of the determinants and, consequently, of the historical provenance of an individual is, namely, a logical impossibility, as Leibniz has demonstrated. An individual becomes individual by way of the totality of the determinants in its history and in the history of the Universe. The infinite number of these determinants cannot be grasped in their totality by a finite spirit. That is, since propositi-

44 Cf. SCHWARTZ (1981).

45 Anyone, who has ever tried to identify unknown fungi with the help of a classification manual, can understand this.

ons concerning individuals are contingent, propositions about species are, consequently, contingent propositions too. For human beings, such propositions are always merely propositions approximating to truth.[46] Individuals are not to be grasped by means of a complete reconstruction of the causal determinants of their existence, because in the causal-mechanical model the following applies: *omne individuum sua tota entitate individuatur* - every individual is made individual through the sum of its being and existence, through its unique *locus* in the co-ordinates of time and space and the framework of causality. The determinants and predicates of the unrepeatable individual are not completely analysable. The factors determining the existence and the entity of the individual are not "solvable" in their causal-mechanical causation.

The individual is then only comprehensible if, alongside the mechanical determinants, an inner power, an entelechy, a form or monad is accepted as being efficacious in the organism. Without the entelechial principle, the unity of form between the members of a species remains incomprehensible.[47] In a reality without entelechy, organisms would be but *petits moments,* points in the Heraclitic stream of reality lacking quality and form. A world consisting of unique points would be, however, like a picture which could have emerged from a radicalisation of the period of art called Pointillism. It would be so pointillist, that no contour-lines or shapes or indeed any thing at all would be perceptible in it. The entelechial principle of the organism has to be thought of together with the uniqueness of the space-time-position of the species as an individual, in order that the organism be recognisable at all.

From this it follows, that the theory of evolution, too, does not have the power to discern and analyse what is real, without the principle of entelechy, an insight, that breaks new trails in the field of biology too: "In place of .... 'Chance and Necessity' - it may be more accurate to characterise evolution as an interactional process that combines the attributes of 'change', necessity and teleonomy".[48] Instead of teleonomy, one might equally say teleology or entelechy. The origin of the species as individuals cannot be reconstructed by evolution theory, i.e. it is not possible to reconstruct the origin of the species by way of selection, adaptation, optimisation or game-theory. What the theory of evolution can effect is a theory of interactions and mutation of the

46 Cf. LEIBNIZ: *Opuscules....*, L. Couturat (ed.), Reprint Hildesheim 1966, 288f. Cf. on Leibniz: RESCHER (1981) and KOSLOWSKI (1985a).

47 Cf. my discussion with Hans Albert on this question in KOSLOWSKI (1985b), p. 74, p. 57, note 26 and p. 58, note 2.

48 CORNING (1981), p. 288.

species that are there, but not a complete reconstruction of the history of what has led to the genesis of the species and to the starting point of the analysis. Moreover, this holds good for social-economic theories also. They, too, are able to analyse only variations of given economic environment, not the economic history as a whole. A reconstruction, which is incomplete, is no re-construction, but a matter of speculation regarding possible worlds.

V. Evolution as History and the Theory of Evolution as a Narrative Theory

Theories about history, even about natural history are possible, but these theories are not causal-analytical theories, which are intersubjectively transmissable in a conclusive manner. They are rather theories that are narrative and historical. If the theory of evolution is expressed as a theory of reality in its entirety and as a theory of *the* evolution - in contrast to the theory about *evolutions* or evolutionary processes - it is a speculative theory like other (speculative) philosophies of history. Its "story" or narrative reconstruction has a certain plausibility, but it also unavoidably contains great gaps, because a reconstruction of reality in its entirety by the causal-deterministic method will always feature "missing links". Missing links are part and parcel of its nature, because a finite consciousness is not capable of reconstructing the historical chain completely .

One such possible history is the hypothesis or précis of evolution as an outcome of a selection, which is purely negative in its effects, in that it "skims off" only the unfit, but lets the fit remain. It is, however, an unlikely story, even a story that appears to be phantastic, that the shapes of the species are supposed to have arisen solely through the erosion of that landscape, which was once formed by the monophyletic life.[49]

This thesis of the differentiation in species by the diversification of the original monophyletic whole is also hard to reconcile with the present general acceptance of the Second Law of Thermodynamics.[50] According to this law, all energy configurations develop into a more probable and less differentiated

---

49 Thus, GHISELIN (1981), p. 282.

50 Cf. on the problem of the reconcilability of the theory of evolution with the Second Law of Thermodynamics WEBER and DEPEW in this volume.

state or form. How can life have arisen as an entropy-absorbing neg-entropic process, i.e. as a constant struggle against the increasing entropy (Schrödinger) from a *monistic* process, which otherwise tends only towards the increase of entropy. Georgescu-Roegen makes the charge against economic theory, that, being so absorbed in the First Law of Thermodynamics and its principle of the conservation of energy, it has forgotten the Second Law and its principle of non-conservation. The same charge must be raised against Darwinian bioeconomics and sociobiology too.[51]

Sociobiology and bioeconomics are new approaches, which expand the Darwinian theory of evolution and transcend the causal-mechanical model of physics with the introduction of economic models, in that they bring in elements of teleological theories of the rational pursuit of ends and of maximisation into the field of biology. In bioeconomics, the behavior of animals is understood as rational, as action under economic laws, analogous to our own self-experience. This development is to be welcomed, since it again emphasises more strongly the oneness of the living between human and other forms of life and might support new approaches to natural law in the human society. The philosophical problem of sociobiology does not rest at the level of the joint analysis of animal and human societies, but at the point where sociobiology makes claims to a holistic, monistic-metaphysical theory of totality. Sociobiology exhibits the extra-scientific character of a philosophy or weltanschauung, not as a theory concerning ecological populations but rather as a theory of the total reality, as a totalistic theory. As evolution-theoretical monism, it is *one* form of *prima philosophia*, one form of metaphysics. In so far as the arguments of sociobiology refer to the entirety of the process of the Universe, they are as little capable of corroboration as they are of being refuted.

Wilson concedes that the theory of evolution as a total theory of the co-evolution of nature and society from the "Big Bang" right down to the modern industrial society is not a scientific hypothesis, but an epic, a story: "The epic story of evolution is as much mythology as the laws (of evolution) are .... a matter of faith, without it being possible to prove them definitively, so that a cause-and-effect continuum from physics to the social sciences would be the outcome". Indeed, it is, according to Wilson, that "the epic of evolution is the best myth, that we shall ever have".[52] By what criteria do we

51 Cf. GEORGESCU-ROEGEN (1978).

52 WILSON (1980), pp. 181 and 189.

decide, which myth is the best? Is there in this question only an arbitrary decision, and are there no rational criteria any more? For the theoretician, Wilson, the answer is unambiguous: the best epic is that, which explains the most.

Even if we were to admit the theory of evolution's greater effectiveness in explaining things, - a matter that is disputed here on account of its problems of incompleteness - the question would remain of whether theoretical proficiency in producing explanations be the sole criterion for coming to a decision between two world views or whether practical and social criteria as well as theoretical ones must both be taken into account. Also, is the theory of evolution more successful in establishing a good life in a just society and in providing an answer to questions about the meaning of life? The theoretical function of world views is only one of their functions, alongside the equally important practical and moral task. It is not theoretical explanations alone, but world views, in which theory and practice unite in a single concept of the right life, that are able to fulfil the four tasks set to world views: to explain the world, to provide motivation for moral action, to overcome the human experience of contingency of existence, and to serve the human self-interpretation, the interpretation of the human as a being with body and soul.

VI. On the Problem of Monism

It is a basic dogma of modern natural science, that science has to be monistic and that only monism is satisfying.[53] The monism or the monistic character of the theory, nevertheless, is not generally carried through at all in the theory of evolution itself. Few evolution theoreticians are so radical as to view the explanation of the evolutionary process also as its justification. Almost all writers in sociobiology are convinced that we are able to turn round the evolutionary process, if we have detected its laws and course, although this possibility to reflect and to turn around the process of evolution implies a sort of freedom of this theory from the process and is difficult to think of in

53 Cf. for much of this GHISELIN (1981), p. 275: "A thoroughgoing Darwinian has to be a monist. There can be no excuse in evolutionary epistemology for Cartesian dualism, Popperian three-world ontologies, or Jamesian pluralistic universes".

a strictly deterministic model. How is it to be explained, that the human species has the power, according to Wilson, to alter its own nature?[54] Why should speech communication liberate humankind from the constraints set by Darwinian principles,[55] if the laws of evolution strictly apply? Liberation and self-transformation in pursuit of purposes set by oneself make no sense in a strongly deterministic or stochastic system. What it means for humans to have a knowledge of themselves and of the process, in which they and their environment are embedded, cannot be comprehended deterministically, because this knowledge assumes a distancing from and an independence of this determination and simultaneously brings it about. The introduction of theory into the system, the duplication of the actuality by way of its conceptual representation and negation, and the repercussions of this image of reality on the shaping of the reality cannot be comprehended adequately, using mechanical-selection-theoretical theories.

The adequate comprehension of the relationships of the sphere of the spirit to that of nature requires a more complex theory than that of causal-mechanical evolution. The relationship of theory and reality, of perception and practice, indicates that a strictly monistic theory of reality in its totality is not possible as a deterministic-mechanistic theory, but, at most, as a theory of the Absolute Spirit, as with Hegel. Monism is only conceivable as monism of the Spirit, in which Nature is conceived of as being the Otherness of the Spirit and the Spirit itself is thought of, speculatively, as that which encompasses itself and its antithesis, Nature.

Deterministic-mechanistic theories, on the other hand, always, of necessity, fall into a latent Cartesian dualism of *res extensa* and *res cogitans*, of Matter and Thought, which makes a consistent monism impossible. The reality, as we experience both it and ourselves in it, is not monistic. It is neither Spirit only, nor Matter only, nor Energy only, nor is it maximisation of genetic fitness. The era of modernity has seen several philosophical and scientistical monisms come and go. Hegel's monism of the Spirit, Haeckel's monism of Matter, Ostwald's monism of energetism and now sociobiological monism encounter one aspect of Reality, but they miss the totality of being, when they take this aspect to be the whole. Monism makes that, which exists, into the night, in which all cats are grey.

---

54 Thus WILSON (1980), p. 196.

55 Thus EIGEN/WINKLER (1975), p. 288.

Every theory of the totality of reality will miss this reality, unless it includes an element of "the affect of admiration and reverence" (F. v. Baader) in the face of this total reality. Every theory of totality must include that there is something left in the total reality that is not explained by monistic theories. All monistic theories that do not include elements of self-restriction will fail since reality in its entirety baffles any attempt to grasp it in one monistic principle.

References

ALCHIAN, A. A.: "Uncertainty, Evolution and Economic Theory", *Journal of Political Economy*, 58 (1950), pp. 211-221.

ALEXANDER, R. D.: "The Evolution of Social Behaviour", *Annual Review of Ecology and Systematics*, 5 (1974), pp. 325-283.

BECKER, G.: "Altruism, Egoism and Genetic Fitness; Economics and Sociobiology", *Journal of Economic Literature,* 14 (1976), pp. 817-826; German text in: *Der Ökonomische Ansatz zur Erklärung menschlichen Verhaltens*, Tübingen (J.C.B. Mohr) 1982, pp. 317-332.

BUCKLEY, W.: "Sociostructural Systems and the Challenge of Socio-biology", in: H. HAKEN (Ed.): *Synergetics. A Workshop,* Berlin (Springer) 1977.

CORNING, P. A.: "Rethinking categories and life", *The Behavioral and Brain Sciences,* 4 (1981), pp. 286-288.

DAWKINS, R.: *Das egoistische Gen,* Berlin (Springer) 1978. Original: *The Selfish Gene,* Oxford (Oxford University Press) 1976.

EIGEN, M., R. WINKLER: *Das Spiel. Naturgesetze steuern den Zufall*, München (Piper) 1975.

FELLMANN, E.: "Darwins Metaphern", *Archiv für Begriffsgeschichte,* 21 (1977), pp. 285-297.

FISHER, R. A.: *The Genetical Theory Of Natural Selection*, Oxford (Oxford University Press) 1930.

GEORGESCU-ROEGEN, N.: "De la Science économique à la bioéconomie", *Revue d'Économie Politique*, 88 (1978), pp. 337-382.

GHISELIN, M. T. (1974a): *"The Economy of Nature and the Evolution of Sex"*, Berkeley (University of California Press) 1974.

GHISELIN, M. T. (1974b): "A Radical Solution to the Species Problem", *Systematic Zoology*, 23 (1974), pp. 536-544.

GHISELIN, M. T.: "Categories, Life and Thinking", *The Behavioral and Brain Sciences*, 4 (1981), pp. 269-313.

GHISELIN, M. T.: Review of "P. Koslowski, *Evolution und Gesellschaft*", *Journal of Social and Biological Structures*, 7 (1984), pp. 391-398.

GOULD, S.: "Evolutionäre Flexibilität und menschliches Bewußtsein", in: P. KOSLOWSKI, R. LÖW, P. KREUZER (Eds.): *Evolution und Freiheit*, Stuttgart (S. Hirzel) 1984, pp. 24-35.

HAMILTON, W. D.: "The Genetical Theory of Social Behaviour (I and II)", *Journal of Theoretical Biology*, 7 (1964), pp. 1-16, 17-32.

HAYEK, F. A. V.: *Die Theorie komplexer Phänomene*, Tübingen (J.C.B. Mohr [Paul Siebeck]) 1972.

HIRSHLEIFER, J.: "Competition, Cooperation and Conflict in Economics and Biology", *American Economic Review*, 68 (1978).

JÜNGER, E.: *Jahre der Okkupation*, Stuttgart (Klett) 1958.

KOSLOWSKI, P.: *Gesellschaft und Staat. Ein unvermeidlicher Dualismus*, Stuttgart (Klett-Cotta) 1982.

KOSLOWSKI, P. (1985a): "Maximum Coordination of Entelechial Individuals. The Metaphysics of Leibniz and Social Philosophy", *Ratio*, 26 (1985), pp. 160-188.

KOSLOWSKI, P. (Ed.) (1985b): *Economics and Philosophy*, Tübingen (J.C.B. Mohr) 1985.

KRINGS, H.: "Sokrates überlebt. Zum Verhältnis von Evolution und Geschichte", in: P. KOSLOWSKI, R. LÖW, P. KREUZER (Eds.): *Evolution und Freiheit*, Stuttgart (S. Hirzel) 1984.

LEWONTIN, R. C.: "Adaptation", *Scientific American*, 239 (1978), pp. 157-169.

LEWONTIN, R. C.: "Fitness, Survival and Optimality", in: DAVID J. HORN, R. D. MITCHELL (Eds.): *Analysis of Ecological Systems*, Columbus (Ohio State University Press) 1979.

LÖW, R.: *Philosophie des Lebendigen. Der Begriff des Organischen bei Kant*, Frankfurt (Suhrkamp) 1980.

LUCE, R. D., H. RAIFFA: *Games and Decisions*, New York (John Wiley) 1957.

LUMSDEN, CH., E. O. WILSON: *Genes, Mind and Culture. The Coevolutionary Process*, Cambridge Mass. (Harvard University Press) 1981.

MARKL, H.: *Aggression und Altruismus. Coevolution der Gegensätze im Sozialverhalten der Tiere*, Konstanz (Konstanzer Universitätsreden Bd. 46) 1976.

MARKL, H. (1980a) (Ed.): *The Evolution of Social Behavior. Hypotheses and Empirical Tests*, Weinheim (Verlag Chemie) 1980.

MARKL, H. (1980b): "Ökologische Grenzen und Evolutionsstrategie Forschung", *Forschung-Mitteilungen der DFG*, 3 (1980).

MASTERS, R. D.: "Sociobiology: science or myth?", *Journal of Social and Biological Structures*, 2 (1979), pp. 245-252.

MAYNARD SMITH, J.: "Evolution and the Theory of Games", *American Scientist*, 64 (1976).

MAYR, E.: *Artbegriff und Evolution*, Hamburg (Parey) 1967. Original: *Animal Species and Evolution*, Cambridge (Harvard University Press) 1963.

MERTON, R. K.: *Social Theory and Social Structure,* Glencoe (Free Press) 1957.

OEHLER, K.: *Ein Mensch zeugt einen Menschen. Über den Mißbrauch der Sprachanalyse in der Aristotelesforschung*, Frankfurt (Klostermann) 1963.

OSTER, G., E. O. WILSON: *Caste and Ecology in the Social Insects*, Princeton (Princeton University Press) 1978, especially Chap. 8: "A Critique of Optimization Theory in Evolutionary Biology".

PETERS, H. M.: "Historische, soziologische und erkenntniskritische Aspekte der Lehre Darwins", in: HANS-GEORG GADAMER, P. VOGLER: *Neue Anthropologie*, Bd. I, Stuttgart (Thieme/dtv) 1972, pp. 326-352.

RAPPORT, D. J., J. E. TURNER: "Economic Models in Ecology", *Science*, 195 (1977), pp. 367-373.

RESCHER, N.: *Leibniz's Metaphysics of Nature,* Dordrecht (Reidel) 1981.

SCHWARTZ, S. P.: "Natural Kinds", *The Behavioral and Brain Sciences*, 4 (1981), pp. 301-302.

SIMON, H. A.: "Rationality as Process and as Product of Thought", *American Economic Review*, 68 (1978), pp. 1-15.

SPAEMANN, R., R. LÖW: *Die Frage wozu? Geschichte und Wiederentdeckung des teleologischen Denkens*, München (Piper) 1981.

THOMAS, W. I.: *Social Behaviour and Personality* (1928), German: *Person und Sozialverhalten*, Neuwied (Luchterhand) 1965.

TRIVERS, R. L.: "The Evolution of Reciprocal Altruism", *Quarterly Review of Biology*, 46 (1971) pp. 35-57.

WILSON, E. O.: *Sociobiology. The New Synthesis*, Cambridge (Harvard University Press) 1975.

WILSON, E. O.: *Biologie als Schicksal. Die soziobiologischen Grundlagen menschlichen Verhaltens*, Berlin (Ullstein) 1980. Original: *On Human Nature*, Cambridge (Harvard University Press) 1978.

List of Authors

SERGEY V. CHEBANOV holds a Ph.D. in Mathematical Linguistics and is Senior Researcher at the Institute of Evolutionary Physiology and Biochemistry, Russian Academy of Sciences, St. Petersburg Branch, St. Petersburg, Russia.

TIM CLUTTON-BROCK is Professor of Animal Ecology at the University of Cambridge, Cambridge, England.

DAVID J. DEPEW is Professor, Department of Communication Studies and the Project on the Rhetoric of Inquiry, University of Iowa, Iowa, USA.

MICHAEL T. GHISELIN is Senior Research Fellow at the Department of Invertebrate Zoology, and Chair of the Center for the History and Philosophy of Science at the California Academy of Sciences, San Francisco, California, USA.

HANS-DIETER GÖRTZ is Professor of Zoology at the University of Stuttgart, Stuttgart, Germany.

HERMANN HAKEN is Professor of Theoretical Physics and Synergetics at the University of Stuttgart, Stuttgart, Germany.

MAE-WAN HO is Professor of Biology at The Open University, Milton Keynes, Buckinghamshire, England.

ALOYS HÜTTERMANN is Professor of Technical Mycology at the Institute of Forest Botany (Forstbotanisches Institut), University of Göttingen, Göttingen, Germany.

PETER KOSLOWSKI is Director of the Forschungsinstitut für Philosophie Hannover and of its Centre for Ethical Economy and Business Culture, and Professor of Philosophy and Political Economy, University of Witten/Herdecke, Germany.

EVA M. NEUMANN-HELD holds a Ph.D. in Biology and is Researcher at the Europäische Akademie zur Erforschung von Folgen wissenschaftlich-technischer Entwicklungen, Bad Neuenahr-Ahrweiler, Germany.

SUSAN OYAMA is Professor of Psychology at the John Jay College of Criminal Justice and at the Department of Psychology of the City University of New York, New York, USA.

LIST OF AUTHORS

GEOFF A. PARKER is Professor of Biology at the University of Liverpool, Liverpool, England.

PETER T. SAUNDERS is Professor of Mathematics at the University of London, London, England.

JACQUES VAUTHIER is Professor of Mathematics at the University of Paris VI (Sorbonne), Paris, France.

BRUCE H. WEBER is Professor of Biochemistry and Chair, Department of Chemistry and Biochemistry, California State University, Fullerton, California, USA.

ULRICH WITT is Professor of Economics at the University of Jena and Director of the Max-Planck-Institute for Research into Economic Systems, Evolutionary Economics Unit, Jena, Germany.

Index of Names

Page numbers in italics refer to quotations in footnotes or references

INDEX OF NAMES

INDEX OF NAMES

INDEX OF NAMES

Druck: Strauss Offsetdruck, Mörlenbach
Verarbeitung: Schäffer, Grünstadt